Steve BACKSHALL

Adventures into undiscovered worlds

EXPEDITION

For Logan

BOOKS

3 5 7 9 10 8 6 4 2

BBC Books, an imprint of Ebury Publishing,
20 Vauxhall Bridge Road,
London SW1V 2SA

BBC Books is part of the Penguin Random House group of companies
whose addresses can be found at global.penguinrandomhouse.com

Penguin
Random House
UK

This book is published to accompany the television series entitled *Expedition
with Steve Backshall* first broadcast on Dave in 2019. Selected episodes were broadcast
on BBC Two under the title *Undiscovered Worlds with Steve Backshall*.

Expedition with Steve Backshall is a True To Nature production.

Executive Producer: Wendy Darke
Series Producers: Susanna Handslip and James Brickell
Series Director: Rosie Gloyns

Illustrations by Ella McLean

First published by BBC Books in 2019
This paperback edition published by BBC Books in 2020

www.penguin.co.uk

A CIP catalogue record for this book is available from the British Library

ISBN 9781785943669

Typeset in 10.08/17.1 pt Bembo Std
by Integra Software Services Pvt. Ltd, Pondicherry

Printed and bound in Great Britain by Clays Ltd, Elcograf S.p.A.

Penguin Random House is committed to a sustainable future for
our business, our readers and our planet. This book is made
from Forest Stewardship Council® certified paper.

MIX
Paper from
responsible sources
FSC® C018179

CONTENTS

CONTENTS

· · · · · · · · ·

INTRODUCTION

My head is filled with white noise. Too-close reverberating roar of falling water, bouncing around the loudspeaker of a rocky room. Crusty eyes open just in time to see a giant centipede scuttle across the rock ceiling, just inches above me. Shudder. Sleep had never come, unkind drips to the face whenever I started to nod off. Beyond me on the narrow shelf we've made home, one of my colleagues was more lucky. Their snoring rumbled in mockery. With limited value in a lie-in I take stock. We have two cereal bars left between us. Fluids will be no problem in the dark roar of the waterfall canyon, but there's no coffee. So no reason to get up. We're sealed in a claustrophobic crack with whitewater cascading down through it. Above us, the nation's highest waterfall. Below, scores more falls and then hundreds of miles of empty jungle. We are, without question, the first human beings ever to venture here. A boulder crashes down into the pool alongside us, rockfall from somewhere high on the vertical cliffs that hem us in. The explosion echoes on and on. Perhaps this place has been unexplored for a reason.

I was almost certain we were going to fail spectacularly.

In 2018–2019, over the course of a single year, my team and I set ourselves the challenge of a fistful of expedition firsts. It was the most ambitious project any of us had attempted, the boldest exped-ition proposal any of us could think of since the voyages of Cousteau's *Calypso*. Our goal was to prove that exploration is not dead, to show that there are still dark, entrancing parts of our planet where no human has ever set foot.

The destinations we chose were the sum of my life's work; entries in a tatty notebook I'd been keeping since my teens. The locations were like treasure maps, with 'X' marking the spot of a fabled cave or annotated with the phone number of a local with privileged access to somewhere special. I'd hoarded these nuggets like golden secrets, more valuable than jewels to anyone in the exploration game.

I once wrote how I always felt as if I had been born in the wrong era. When I was younger, I yearned to have been around when explorers could have sailed off the edge of the earth, when they genuinely had something grand to discover for the first time: continents and mountain ranges, islands and oases. These golden-age explorers really had something to risk too – usually their lives, or the lives of their crew, consumed whole by mythical giant snakes, eaten by unfriendly locals, or lost to parasites that ate them from the inside out.

Scary as all of this sounded, as a kid I felt a paralyzing disap-pointment that I was born too late, in a generation when all of the proper discoveries had already been made (though admittedly I didn't mind being safe from smallpox and syphilis). Now, the real exploration was being made in labs, through genetics and genomics and at the sharp end of electron scanning microscopes. Nobody in my era would ever be the first to step out of the desert and look onto the Grand Canyon. No one would ever again discover the source of the Nile, be the first to summit Everest, make the first giant leap onto the surface of the moon.

However, after spending most of my adult life immersed in the field of old-fashioned expeditions, I was beginning to realize that my pessimism had been unjustified. In fact, if anything, we were on the cusp of a new golden era of exploration. The technology that seemed to make the world painfully small has also driven expeditions into a new realm – an era of ambition and information-sharing fuelled by satellite mapping, drones and other space-age technology. In fact, you couldn't even call it 'the Space Age' any more – it was iPhone, graphene, Higgs boson, Dark Matter-Age tech!

And it turned out there were still old-fashioned expeditions to be had of grand scale and scope. My friend Ed Stafford became the first person ever to walk the entire Amazon river in 2009. That same year, Hang Son Doong cave in Vietnam, the largest in the world, was surveyed for the first time. Felix Baumgartner took a high dive from space and we're talking seriously about travelling to Mars. Our deep oceans are less known than the surface of the moon and spit out bizarre aquatic oddities daily, and should you smoke out any rainforest tree, you'll unquestionably find dozens of species that are new to science. *National Geographic's* yellow-framed magazine cover still screams monthly of some mighty conquest or discovery . . .

Exploration wasn't dead, it just needed a little more creativity. Adventure is more accessible now, with more focused experts, more technology and more understanding than at any other time in human history.

The time was ripe for a major compendium of modern exploration to take place – but it had to be done now, as an ever-hungrier human populace raced across the globe seeking firsts for themselves. That, however, was easier said than done. We might be certain about how spectacular a mega transect of the globe could be, but the only way to fund it would be to convince a lot of powerful people to put their trust in me and my team. The great Ernest Shackleton wrote in his memoirs of how the greatest slog in expeditions is managing to get them funded. He relied on both grants and private funding; his major donors were immortalized in the names of his lifeboats: James Caird, Dudley Docker, Stancomb Wills. For me it would be a different enterprise.

The pitch consisted of a list of big firsts that were out there still to be done. Genuine ones, not carefully worded false firsts: 'the first British male over forty to climb K2's south face without oxygen', or 'the first Scottish schoolboy to paddle across the Atlantic in a bath tub while wearing a gorilla suit and a tutu'. No. These would all be dramatic and iconic journeys we could really get excited about. And it wouldn't just be us that would get excited.

Just imagine how many kids there must be all over the world who are like I was. Crouched with a flying fist over their Xboxes, but actually filled with a deep existential ennui, frozen by their disappointment. Wannabe Amelia Earharts and George Mallorys, fledgling wanderlings who just need a little inspiration in the form of proof. A grand, multi-faceted expedition could convince them that there is a world out there still left to explore. It could enthuse the next wave of archaeologists, zoologists, conservationists. Just the kind of thing the teenage me would have eaten up!

For 20 years, every capable contact I ever met I'd grill, in search of secret sites and possible leads. I kept a filing cabinet filled with folders marked 'Yamal Peninsula – Mammoth carcasses and dog sledding tribes', 'Aleutians – first open sea crossing', and 'Kuril Islands by kayak' – all stuffed with photos and articles, email addresses and business cards. None of the projects could be found on Google. Some of the rivers, caves and peaks couldn't even be seen in high-resolution satellite imagery.

Some came and went. I spent my savings planning the first ever sea kayak circumnavigation of Svalbard, only to be let down by sponsors. I watched on tenterhooks while a Norwegian team attempted the same mission, though they had to abandon theirs after one team member was dragged out of his tent by a polar bear. But finally it was completed by a heroic three-person team. I saluted their accomplishment, while secretly cursing them under my breath.

The first expedition I ever put in my folder, however, stayed untouched and unblemished, until finally it became a reality. From 1997, when I first went there as a writer, I desperately coveted the first ever descent of the Baliem River in Papua (the western half of New Guinea). It consisted of 500 miles of rampaging white water, dropping from the towering central highlands all the way to the coast, across some of the most scintillating tribal lands on earth. The whitewater would be first thrilling, then beyond dangerous, before the river descended into the darkest, deepest malarial jungles, and on through giant croc-filled estuaries and swamps to the sea. It was perfect.

I started trying to get the Baliem expedition off the ground in 1998, and not a month went by that I wouldn't go cap in hand, trying to convince businesses, wealthy patrons and TV executives that they should commission me to do it. Every single boss I ever had while working in television got the pitch over and over until they told me not to mention it again.

Finally, in 2016, the opportunity arose, and we attempted the river, 18 long years after my first journey there. The result was far from flawless; we spent more time sitting bartering with local people than paddling, and ended up having to do the section to the sea by motor boat. Even with huge delays in the field,* the journey yielded

---

* Soon after arriving in the country, our military advisor got word that freedom fighters were coming out of the mountains armed with AK47s to take us hostage. This had happened to another team a few years previously and ended really badly, with several fatalities. We had no choice but to skip several sections of the river, and stay beyond the grasp of our kidnappers. This, combined

a successful television program that sold all round the world, and recouped its investments. It provided some evidence that we could get our grand expedition project to succeed. It would, though, come with numerous caveats. We'd need to film the entire enterprise, which meant self-filming much of it and bringing along film crews for other parts. For boring reasons of finance, we'd also need to complete the entire project in 12 months, which was insanity. But we couldn't say no; this opportunity would only come around once.

The first bottle of celebratory champagne was sipped with caution. I'm pretty sure that not a single one of our fledgling team actually believed it was possible.

Assembling the full team was going to be critical. Our goal was to bring together the finest climbers, cavers, divers, paddlers and jungle jocks, while taking care to avoid the monster egos that can accompany great ability and achievement. Some choices were obvious, as I'd done big trips with them in the past. Newer team members were more of a gamble. At the very least, someone who snores or stashes all the best food for themselves could fray morale. At worst, a teammate who can't be counted on could let go of a rope with you at the other end of it.

In the planning stages, the team was based out of Bristol, with

----

with several illnesses on the team, plus unseasonably high rains and flood-condition waters, meant we were far from completing the whole source-to-sea descent.

about ten hard-nosed adventure bunnies all crammed together shoulder to shoulder in the attic rooms of a busy production house. Here, they worked tirelessly for a year, organizing flights to places that no one's ever heard of, and expeditions to places that don't exist on any maps. The walls were plastered in plane schedules, spreadsheets, calendars and charts, all scribbled over in highlighter pen and covered in Post-it notes.

For every inch of each journey, we needed worst-case scenario exit and evacuation strategies, in case of broken bones, snakebites, illness and worse. Evacuation protocols were sometimes at best a fantasy. The unwritten part of every risk assessment was the bit no one wanted to even think about: body recovery.

Things certainly didn't go without a hitch. We had to pull out of our Colombia mission just days before leaving, as local cartels and guerrillas refused to guarantee us safe passage in dangerous territory. At least one person got ill on every single expedition – thankfully, though, never seriously. Rockfall nearly cost us the ultimate price on three separate occasions, and looking back, we were really lucky to have escaped unscathed.

But nobody got hurt, deported or ended up languishing in a foreign jail, and none of the expeditions failed. In 12 months we made ten trips to fresh destinations around the globe. They included first descents of three rivers, two of which didn't even exist on any maps. In a time when running a single new rapid can be a big deal, we thrashed down several hundred unmapped, unknown and often unexpected cataracts. We took light into cave systems that have never been illuminated, and mapped over a kilometre of sunken cave passage hidden beneath the limestone of the Yucatan, also discovering a dazzling Mayan underworld. The first ascent of an unnamed Arctic mountain was followed by the exploration of an unnamed desert canyon. We discovered a waterfall that even satellites hadn't picked up, and is one of the highest in its nation. We climbed a stunning route up a mighty unattempted desert cliff face, and time and time again we were able to say something along the lines of, 'no human being has ever stood here before'.

During the same year, I spent my life savings on building a house, flirted with bankruptcy, held my first child and said good-bye to another. I came within seconds of losing my life and was saved by a new friend. There were moments of indescribable ecstasy offset by crushing drudgery, terror alongside boredom and joy mixed with pain. There were perfect pristine places that gave me hope for the future of our planet, and hellholes that left me thinking we're all totally screwed and might as well give up. We had close calls and near misses, came nose to snout with a hungry polar bear and the world's most dangerous snake. We lost old friends, and forged fast friendships with new ones.

It was a hyper-real experience, and one that taught me a dozen valuable lessons. It showed me that the efforts of a hundred people is nothing compared to one exceptional friend who's really got your back. That the world's greatest IMAX theatre can't compare to a roaring fire beneath the Milky Way, shooting stars coursing across a desert night. That everything tastes better outside when you're hungry (though even starvation can't make a freeze-dried diet good). These secret sites we discovered, these remote and wild corners of our planet, offered more succour for the soul than all the world's cities combined.

It was the most surreal and dramatic year imaginable, and cramming it all into a single book has been next to impossible.

1. UNDERWORLD

MEXICO

Fangs, tusks and terrible canine teeth line the maw of the giant beast, beyond it a chilling gullet of ink-blue nothing. The gap between the stalactite teeth is cadaver-wide; the only way to penetrate the guts of the gorgon is to roll onto my side and edge through, inch by inch. My diving tanks thong and ring out on the rock, resounding like the toll of a Buddhist bell. Light from my companions slices in shafts, beckoning me on. I twist, roll and creep, easing myself forward into danger. Every breath bubbles to the ceiling, where it turns to quicksilver, spreading and slithering its mercury molten metal glaze across the roof like a living thing. And then a catch, a set of hidden claws flexes to snag me, holding me back. The urge to panic seethes red from within, my breathing hastens, my heart rate races. Above me the weight of thousands of tonnes of oblivious limestone seems to descend, like in one of those tacky old Superman movies where the walls of the villain's lair close in to squish our hero flat. The only way is on, slowly and calmly, further from air, light and help with every second.

Ask any American what they know of the Yucatan, and they'll say, 'Cancun, right? That place where all the drunk kids go for spring break?' I doubt anyone, anywhere in the world, would go, 'Oh yeah, the Yucatan? That place is Ground Zero for exploration!' Millions of tourists pour through this little corner of Mexico every year, little knowing that they may be mere miles away from places humans have never been, many of those locations heart-stoppingly beautiful.

The Yucatan peninsula is a low, flat shelf separating the Caribbean Sea from the Gulf of Mexico, and is a stronghold of modern Mayan culture. The peninsula is an extraordinary piece of geology – arguably the most significant on planet earth. It forms the southern crater rim of Chicxulub, the site of the asteroid impact that led to the extinction of most of life on earth around 65 million years ago. The six-mile-wide hunk of space junk landed in a shallow sea, throwing not just water but sea-bed ejecta into the atmosphere, which circulated the planet for many years, blotting out the sun and causing a 'nuclear winter' that eradicated the non-avian dinosaurs.*

As if this cataclysmic impact wasn't enough, the Yucatan peninsula's unique topography has been formed by millennia of rising

---

* If the asteroid had struck just a short distance further out to sea – which was statistically more likely, as the deep sea is the biggest biome on the planet – the localized tsunamis would have been catastrophic, but the atmosphere would not have filled with choking dust, and we might be living alongside Gigantoraptors and Brachiosauruses even today.

and falling sea levels, leaving behind them bizarre landforms. Tens of millions of years ago, the entire region was underwater, and around the edge of the Chicxulub crater rim a vast coral reef formed, the limestone skeletons of coral creatures forming a solid structure that must have rivalled the modern Great Barrier Reef for abundance.

Then, during the last ice age, much of the world's seawater became steadily bound up in ice. With more water frozen, the sea level dropped dramatically and the coral limestone was exposed to the air – the opposite of what we're facing today due to human-induced climate change. The reefs dried out, and in this new terrestrial time, limestone did what limestone does. Almost all caves around the world are formed in limestone, not just from erosion, but from the acidic effect carbon dioxide has when dissolved in water. Acidic fluids started to permeate the rock, fizzling and burrowing, while simultaneously depositing extravagant speleothems such as stalagmites and stalactites throughout the caverns. This process took tens of thousands of years, but this was a mere mermaid's fart in geological time, a blink of an obscure deity's eye. The end result was the formation of some of the most extensive and gloriously decorated cave systems our planet has ever seen: thousands of miles of grottoes and caverns, an underworld that would dwarf Tolkien's Mines of Moria.

The young post-Paleozoic limestone was porous without much soil on top, so hardly any standing water and no running water could remain on the surface in Yucatan, and in the driest of dry seasons none at all. Instead, it filtered down through the limestone until it reached saturation, forming a water table. The latticework glugged full of water, and the caves and everything within them was flooded and frozen in time.

That wasn't the end of the story, though. The landscape continues to evolve, to erode and in some places to grow. The massive caverns can't always withstand the weight of their own rock ceilings. In times when water levels are too low to float-support the roof above, these roofs can collapse, creating sinkholes.

Cenote, Ts'onot or D'znot are Mayan words which mean simply 'cave filled with water'.* The Mayans gave enormous significance to these cenotes. Not only were they their only source of fresh water, but they were also seen as gateways to another world, a place that gave life and held unimaginable treasures. Cenotes were a key part of the success of the ancient Maya from around 1800 BC to AD 250, irrigating their crops, slaking the thirst of their growing population, enabling the Maya to grow into one of the great civilizations of the times, with accomplished engineering, astronomy and mathematics.

Within the great Mayan city of Chichen Itza was a sacred cenote that became its most important shrine. It was named the 'Cenote of Sacrifice' by Spanish historians as the Mayans so regularly sacrificed people there, throwing their bodies into the flooded

---

* Which is frankly a little disappointing. I was hoping for something more poetic, like 'eye to the underworld'.

holes. Human offerings were mostly male, often enemies, stolen children or those who were older and without dependents. As well as unfortunate people, members of the Cult of the Cenote threw in obsidian and jade, as well as gold artefacts from Panama and Costa Rica. It's thought that they were offerings made in times of drought to the rain god Chaac, which I guess makes sense, praying for rain at the temple where life-giving water emerges.

From 1904 to 1911 archaeologist Edward Thompson dredged and eventually dived here, removing 30,000 ceremonial pieces from this one sacred cenote. These items documented eight centuries of an entire culture near expunged by the Spanish conquistadores. This treasure haul was more than a historian could dream of, riches beyond the ken of men. With this in mind, it's extraordinary that so many of the Cenotes remain unexplored to this day.

This is where *Indiana Jones* should have been set.

The Cenotes first came to my attention in 2009, while working in the Yucatan peninsula. One of the side effects of Scuba diving in sinkholes with open ceilings is that at certain times of the day sunlight cuts down through the waters like Broadway spotlights, as if the gods have blessed the depths with celestial spells. The water was rainwater that had percolated slowly down through the rock, and thus had been filtered clean of all its suspended detritus. It was purer than Peckham Spring; so clear you could rarely see it was water.

But beautiful as they were, the cenotes I saw were well dived. Many of them had concrete steps leading down to them, and car parks just metres away. With the spring break haven of Cancun only two hours' drive away, they were more like a scuba divers' theme park than a site for serious exploration. But with my interest piqued I started to do a little digging. (Not literally, I'll leave that to the British cavers.)

Apparently, there were at least 6,000 cenotes throughout Yucatan, and yet no more than 40 or 50 were ever dived. Cave diving is such a potentially dangerous activity that once you penetrate beyond the light you need to dive with an immense amount of back-up or redundancy. So, as well as your standard kit, you need at least two scuba cylinders for every dive, and often four or more, plus dozens of lights. To stagger even a couple of hundred metres with all that equipment is really tough going, so most divers just stuck to the 'drive-in' dives.

Even within those easily accessed cave systems were passages that were without question not explored. Dos Ojos (the two eyes) is one of the best known of all the cenotes, yet in 2018 a completely new passageway was revealed, linking it up to another system and making it the biggest flooded cave passage on the planet . . . and Dos Ojos is dived by several hundred people a day. Go a stone's throw further into the forest, and there were cenotes that had only been dived a handful of times.

My 2009 trip concluded, and as I flew out of Tulum airport, I looked down. Forest. Miles and miles and miles of forest. Despite the fact that the Yucatan is Mexico's best-known tourist destination, there was just one coastal road, and then inland of that was hundreds of square miles of brutal, scratchy, horsefly-ridden, dry forest that ran all the way to Belize in the south and covered three huge Mexican states. And that impenetrable scrub overlaid the same unique geology: limestone bedrock uniformly riddled with cave passages, most of which would be filled with water. If the caves within minutes' walk of the road were not even fully explored, imagine what you'd find if you walked in for an hour. Or a day. Or if you were to drop into the middle of the forest by helicopter. You'd not just find new cave passages, you'd be pretty much guaranteed to find whole cenotes that no one knew of. And what would be inside them? Mayan temples? Human remains? Gold, jade, obsidian (whatever the hell obsidian is)? My head swirling with images of museums filled with treasures I had collected, I started to scribble a plan.

Although I'd be inspired by the potential of the cenotes, I was soon distracted by other projects and it wasn't until my honeymoon in 2016 that my hunger for them returned with a vengeance. My great friend, fixer and hardcore diver Scott Carnahan had arranged for my wife Helen and me to spend a week on a boat he looked after in Mexico's Sea of Cortez.

Scott not only looks like Jesus, with dark hair that cascades down his spine and luxuriant beard, but he brings divine blessings too. During our blissful time amongst the islands and wildlife of that true blue wonderland, Scott pulled out a recent copy of *National Geographic* he'd been saving for me. The front cover showed an image of a diver within a sunken cave system surrounded by deep blue, and in front of him, illuminated by his torchlight, was the skull of an extinct mammoth-like elephant. The *Nat Geo* team's exploration of some of the Yucatan's cenotes had revealed prehistoric animal bones within the sunken cave systems. Massive intact skulls of mastodons, ground sloth and cave bears. These wonders were in systems that also held human skulls from Mayan ritual sacrifice. These discoveries would have been beyond exciting wherever they were found, but diving? Alongside perfect stalactites and stalagmites in water so clear it looked as if the explorers were flying?! It was the stuff of pure Hollywood fantasy. I suggested that Scott should start assembling our very own Aquatic Avengers for a voyage into the Underworld. 'That certainly wouldn't suck,' Scott said. He says this quite a lot.

But like a fortune cookie, life sometimes delivers a sweet crunch, followed by something simple and profound that could make you re-evaluate everything. Helen and I were expecting We were cautious, and left it to 20 weeks to tell the world.

By then, we knew the twins by name. Our life was mapped out ahead of us; Helen would, as always, be superhuman through pregnancy and their young childhood, and would take a private coach so she could return to rowing – perhaps even in time to make the Tokyo Olympics. We would bring the boys up to be athletes, adventurers, naturalists . . .

We planned every day, brimming with excitement. And then, at 23 weeks, we found that one of the boys hadn't made it.

Helen was now a high-risk pregnancy, with the potential of going into early labour. The consultant advised against her travelling more than an hour away from a specialist neonatal unit, which meant she couldn't accompany me on any of the trips. My first two back-to-back expeditions totalled ten weeks away. We had planned them that way to enable me to get back in time for the birth, and then to have six weeks at home to learn how to be a daddy. But now the *Expedition* team was committed, everything was organized.

There were tough times before that first trip. Helen and I talked seriously about whether I should go at all. There was a risk of things starting to happen while I was underground in Mexico or out of contact in the Arctic. On the other hand, if I stayed, the whole *Expedition* project would be over.

The expeditions that had seemed the most important thing in the world suddenly seemed meaningless. Helen was stronger than me, and insisted I go. 'People in the military do this all the time,' she said. For the next two weeks we tried to freeze and frame every second together. Instead of excitement, for the first time in a quarter of a century travelling for a living, I got on the plane filled with doubt and regret.

Often when you're putting together an expedition team, you find there are only one or two people in the world that can do the job, and all the logistics revolve around them. Scott Carnahan is one of those people. He began to put together a list of names of people with phenomenal talent and credentials.

The first name on the team sheet was Robbie Schmittner. Robbie was without question the finest cave diving explorer in Yucatan, and by common consent of everyone who worked there, our golden ticket to finding something new and special in the Underworld. He was an arresting sight, his hair shaved into a protégé Mohawk, sunburnt, and with brilliant blue eyes commanding your attention. German by birth, but with five children all in the Yucatan, Robbie had spent 20 years exploring underground in Mexico. Arguably no one in human history has uncovered more lost darkness than Robbie – at least 300 kilometres (imagine swimming from London to Manchester. Underground).

Robbie was fresh from the glory of discovering a through passage that linked up two vast cave systems in the Tulum area. The connection was one he had been seeking for 15 years, and brought the ultimate glory – discoverer of the aforementioned largest cave system on earth. He told us of the battle to find the connection, narrowing down hundreds of kilometres of caves using sophisticated mapping software to try and find the spot where the two systems came closest to each other. Then months of diving the same passages over and over again, searching for signs of current, for tiny cracks that hadn't been investigated. Finally, at the end of a six-hour dive, he found a sliver of rock with water rushing through it. Without the air to try and dig through, he pushed three metres of line into the current, and it was carried beyond into the unknown. He then came at the constriction from the other end, again spending many hours underwater to get close to where he thought the join might be.

After days of divining and frustration, he found a tiny slit that he must have swum past a hundred times. When he put his hand over it, he felt water pumping through with huge power. Squeezing himself into the gap, he craned his neck around and saw the tatty end of his line flapping in the flow like a sock on a windy washing line. With that, a cave system 600km long was christened. And that must surely rate amongst the great achievements of discovery of our times.

I watched dozens of videos of Robbie's explorations online, most of them with my jaw on the floor. The audacity of his dives, the scale of them, was almost impossible to compute. Dives lasting seven or eight hours, with multiple tanks littered throughout cave systems enabling him to penetrate sunken caverns that you could comfortably park a fleet of jumbo jets in. The fact that Scott rated him the best in the world would have been enough for me to trust him, but within minutes of meeting him I also really took to him as a person. Then my job was to prove to Robbie that, despite my somewhat rough diving skills, I wasn't just some hopeless tourist.

Our cave camera operator was Katy Fraser, recommended by an old diving contact. She was only 24. It was tricky trying to convince the team that instead of hiring the most experienced cave camera operator on earth we should place our faith in an unknown, but fortunately the team agreed to take the gamble.

Bernadette Carrion was the first female cave dive instructor in Mexico. She is the kind of person that anyone and everyone would instantly take to. She's tiny, coming up to my armpit, with a naughty twinkle in her dark eyes and an easy manner, but she also has a set to her jaw that says she is much tougher than first appearances might suggest. She would have a vital role to play, as a guide to the unique environments of the cenotes, to the Mayan mythology, and to the whole method of diving in these unique and fragile environments.

The first night I met Katy and Bernadette was at a beach bar in Tulum. It was immediately obvious that they had no idea I'd been diving for 25 years, had my commercial diver's licence and was fully cave qualified. In fact, they thought I'd only learned how to dive for this trip. Bernadette offered to take me out the following day to teach me how to kick my fins, and Katy reassured me, to prepare me for the first moment I swam into a cave, that 'the lights can be quite blinding'. She explained, 'Your first instinct will be to panic, but we'll look after you.' I tried dropping subtle stories about all the cave diving I'd done, but they didn't catch the hint.

A little later, Katy talked about how she would have to wear a much thicker wetsuit than me: 'You have much more bioprene than me,' she confided seriously. She was calling me fat. For the rest of the evening I found myself sucking in my stomach. She also repeatedly referred to people from 'my generation'. When I told James, my esteemed long-term director, he said, 'That's nothing. I went to lift a chair yesterday, and she came over and asked if she could do it for me. She wasn't being polite, she just honestly thought I needed the help.'

Still, it must have taken something pretty special for a girl from a posh private school in England to bin university and come out to Mexico on her own to take up a life filming people diving in caves. Katy had spent years and years inside Yucatan's caves, making them and their divers look beautiful. We knew she would work hard and bring a craftsperson's eye to the visuals. We demonstrated the immense respect we had for her work by christening her 'Gap Year' and constantly mansplaining simple tasks to her. (After a few days on expeditions, the humour devolves straight to the playground.)

Though the main expedition team was set to be in the Yucatan for just five weeks, I arrived a week early in order to train specific-ally with Robbie, Katy and Bernadette – perfecting our systems and underwater communication so we would all be aligned when putting ourselves in harm's way.

Serious cave diving is very different in its techniques to con-ventional open water scuba. This is mostly down to the fact that the consequences are far more severe if anything goes wrong. Running out of light, getting lost and, worst of all, running out of air *will* result in a fatality. In the open sea, even if you're quite deep, if your systems fail then at least you can bail out and head for the surface. In cave, there is no surface. Cavers dive on at least two separate tanks/cylinders, and work to a 'rule of thirds'. This means that you work out how much gas you have available to breathe and divide into thirds: one third to head in, one third to head back, one third left for redundancy and emergencies. This sounds super cautious, until you figure that if one system com-pletely fails at the furthest extent of your dive, then your other tank will have exactly the right amount of air to get back to the light. Exactly. Anything else goes wrong – breathe too hard because you're scared, take a wrong turn, have any other system failures – and you will die.

Say it like that, and it sounds horrific. For the most part, it is just an absolute joy. Our first few days were terrific, plunging into a range of well-known Yucatan cenotes within half an hour of the towns of Tulum and Playa del Carmen. These had exotic Mayan names like Xunan Ha and Tajma Ha, Spanish names like Angelita, Casa Cenote and Dos Pijos, and British names like Dream-gate and Carwash – the latter named because people once used the water to wash their trucks. You could literally pull up on the side of the highway, reverse your truck up to the side of the cenote, and within 30 seconds be splashing round in clear waters full of terrapins, aquarium fish and even a baby crocodile.

All the training dives went beautifully, except one. Angelita – the little Angel – was to have a bite to her that was more devil than deity. This cenote is pretty anonymous from the surface, but below is remarkable. It's a vertical sinkhole that drops straight down to 80 metres in depth. However, over time the centre of the cenote has filled up with organic junk, making a hill in the middle of it that comes up to 30 metres from the surface. Above the 'hill' is fresh water, and below is salty. A gap between the two is known as a halocline – halo means salt – and they are common in cenotes. Here, however, the rotting organic material has created gas hydrogen sulphide which sits in a sinister underwater cloud over the halocline. I'd never seen anything like it. However, my first sight was very nearly my last.

Our goal for the morning was to freedive down to the hydrogen sulphide layer at 30 metres. Freediving would unquestionably be a vital part of our exploration, and this would form a perfect training scenario. We sent Katy, Bernie and photographer Gabriel down on scuba to light up the spectacular hydrogen sulphide mist, and wait for me to freedive down to them. In the meantime, I lay at the surface relaxing, bringing my heart rate down, and 'packing' – breathing deeply to oxygenate my blood, and turn my muscles into my own scuba tank. I'd been training at this for several months beforehand, doing freediving breathhold practice in the pool, in the car, even while writing on my laptop, and had managed to get my static apnoea (just holding your breath without swimming or moving at all) up to a maximum of five minutes. Down to 30 metres and back up should not take much longer than one and a half minutes. However, all was not well. There were three or four teams of other divers around, plus a team of four freedivers at the surface. There was also a thick green layer of algae-rich water that extended from the surface down to about 14 metres, so once the team had descended, I could only guess where they were.

In retrospect, we should have just called off the dive; too much was out of our control, and there was too much to potentially go wrong. However, I had my mind on the task of filling my red blood cells with lovely oxygen, and was totally in the zone. Bang on my time cue, I swam a few strokes forward, and dived down into the green. With my weight belt on and huge long freediving fins, I dropped down like a rock. Fifteen metres, twenty, then I popped out of the gloom and was greeted with a sight of utter wonder. The gas layer hung like fog over a midnight phantom swamp, ancient tree branches poking up from the cloud like the masts of listing ghost ships. Around them were the dark cauldron walls, the mist like a bubbling potion beneath. I would have gasped, if my mouth wasn't tight shut.

But then I realized my mistake – I had been swimming towards the wrong dive lights. My own team was further away than I expected. I swam laterally towards them, burning precious oxygen with every millisecond my fins were pumping, leg muscles firing, filling my body with lactic acid and carbon dioxide. I reached them, but had totally overstretched myself. I needed to be at the surface *now*, and it was 30 metres above me.

I fought to keep from panicking. I must not kick hard, that would only burn more oxygen and I would never make it. Robbie, my safety diver, was just below the surface; I had to at least get to him. I tried to swim up as steadily as possible, but my throat was contorting, battling to make me breathe. If I gave in to the temptation and sucked in, my lungs would fill with water and I would drown. My little training dive had turned into a serious situation.

But now I could see the surface, reflected light with yellow leaves dancing in circles. It took an eternity, battling against my dive weights, more desperate to breathe than ever before. I burst through the surface, and then everything was a blur. It felt as if my whole body started convulsing, I was choking, gasping, and then a strong hand grabbed me.

The next thing I knew I was at the landing stage. And someone was shouting at me. When the fuzz cleared I realized what had happened. I had been overwhelmed with what freedivers call a 'shallow water blackout'. This occurs when carbon dioxide builds up and, starved of oxygen, your brain shuts you down. Bizarrely, it most often happens at the surface, when you take your first breath. But why the hell was someone shouting abuse at me? It was one of the freedivers, and he was not being polite. 'I saved your life,' he shouted, with a heavy French accent. 'What the hell are you doing freediving here, you have no business here, where is your safety diver?' I motioned quizzically to Scott and to Robbie. They were there, ready to save me, but the French guy had got there first. I thanked him humbly and not a little ashamed. No one likes to be told off. And particularly not when they really have unquestionably messed up.

I was a bit shaken, and certainly wasn't going to be doing another freedive anytime soon.

That evening we all gathered in a thatched-rooftop hacienda, clustered round maps and Robbie's laptop, and started hatching plans for the mission ahead. Robbie had an area in mind where there were at least two or three entrances to a cenote that he knew of but had never dived. I felt a prickle of anticipation up my spine as he teased us with the kind of things he expected to find. Exactly as we began talking about heading into new caves, the sky started to flash about us, and then a biblical downpour thundered down from the heavens. The lightning was so intense it seemed it would split the sky.

'The ancient Maya used to make sacrifices to the rain gods in the cenotes,' Bernie said. 'They'd take a child and rip its heart out, still beating, as sacrifice.'

'When we're exploring, we at least need permission from local spirit men for sure,' Scott added.

'What do you know?' I said. 'You told me it was the dry season.'

'Perhaps we need to sacrifice one of us,' Robbie joked, looking up at the rampaging heavens, 'Katy, you're the youngest . . .'

We all chuckled, but there's no doubting there was a strange energy about the whole evening. Maybe it's nothing; people often talk about feeling on edge during tropical electrical storms. Plus, I was spending too much time reading endless research texts on Mayan human sacrifice, while spending my days floating through bizarre underworlds . . . but even so. Even for a cynical old science nerd like me, it all felt a bit odd. Could the gods really be letting us know they didn't want the Underworld explored?

After my embarrassment at Angelita, we needed to get away from the 'Disney cenotes', where hundreds of tourists were splashing around in the shallows. So Robbie took us somewhere a bit more off the beaten track. We rattled into the forest down a 4X4 track for a couple of miles, before shouldering our scuba tanks and taking to the jungle on foot. We didn't have to go far, following the calls of the motmot birds towards the water.

The turquoise-browed motmot is the most important creature of the cenote environment. Two barbs extend from its blue tail, which conclude in a distinctive shimmering pendant. My book, *Animals of the Ancient Maya,* says these have 'bald notches where the bird has plucked the feathers away with their serrated bill,'[1] though this seemed pretty unlikely, there are plenty of sources suggesting the connections of the feathers are weak, and they get lost through preening. The ancient Maya believed the motmots were birds of medicine men and the brave, and they are eternally linked to the cenotes. They choose to nest within the cave opening and are omnipresent around the cenotes, and indeed can be a tantalizing way of finding caves that might be near invisible from the surface. The national bird of both Nicaragua and El Salvador, the motmot is known here in Yucatan as the 'clock bird' due to its habit of ticking its tail from side to side while perched, with the disk at the end swinging like a pendulum.

This is about as close as Central America comes to a bird-of-paradise, though unlike the birds-of-paradise both males and females display the bizarre elongated tail. Potentially the two sexes use their tails in different ways. For the males, it is pure sexual selection; the tail is a classic piece of 'honest signalling', a handicap that cannot be faked. 'Look at my flamboyant tail,' he seems to say, 'imagine how difficult it is for me to get around with this dragging behind me, imagine how good my genes must be for me to have even made it to adulthood!' In classic sexual selection, the female bird would then be dowdy and camouflaged, but not in the motmot. Here, she is equally coloured, so there must be another function to the display. Perhaps it is a 'pursuit deterrent signal', a message to predators that the bird must be strong to risk standing out so much, and thus is simply not worth pursuing.

The motmot was an important bird to the Maya, as it leads the way to cenotes and thus life-giving water, and so they had many legends about its presence. Some said that all the birds had come together to build a shelter during a mighty storm, but the motmot was too arrogant and refused to help. When the storm hit, it was stranded with its tail out in the open, and the rain stripped it of its luxuriant feathers. Another says that a god stamped on its tail and ripped the feathers clean away.

The cenotes were alive with them, with sometimes four or five clustered on a single branch hanging over the water, calling out their own name as so many birds here do. (You should hear the chachalaca in full voice!) Birds in fact were one of the great wonders of the dry forest that surrounded the cenotes. On my second day I watched in stunned silence as a pair of mini toucans perched in the treetop beside me as I ate my breakfast. These collared aracari were no bigger than blackbirds, yet carried dark bills the size of slightly flattened overripe bananas. Magnificent and piratical frigate birds loomed overhead at the coast, as did high-diving plummeting pelicans.

Perhaps the most conspicuous bird was the great-tailed grackle, which sounds like a mythical creature that should be stealing Hermione Granger's handbag. Grackles act and look a bit like crows: adaptable, noisy, black and bold, able to tackle the same simple problems as corvids (the crow family) in other parts of the world. Which would seem cause to celebrate them . . . except they are really, really loud! The Mayans said grackles have seven songs that reflect the seven great human passions of love, hate, fear, courage, joy, anger and sadness. Which is charitable, to say the least. Try sleeping with an incessantly squawking grackle outside your bedroom window when you're jetlagged; I reckon you could add 'homicidal' (avi-cidal?!) to the list of passions. And while we're on the subject, 'songs'? I've heard every sound coming out of them other than a song, from sexist wolf whistles to the clearing of a throat prior to spitting a whopping greeny to tuning an old-timey radio, and caws that sound like a cat's claws running down a blackboard. Yet for all their tone-deaf crooning, the grackle is a bird that does really well around humans, scoffing bugs off parked car number plates, and the French fries from your plate if you're not looking. Over the last hundred years they have moved all the way from South to North America, expanding their range by moving along the growing human corridors between the two. Now in Texas, they gather in crowds of thousands, and are known as 'tacoraptors' and 'horny devil birds'. It seems once we manage to nuke ourselves to death, all that will survive are cockroaches, rats and grackles.

Tanks bruising our shoulders, we carried on down the track,

wandering past a forgotten Mayan ruin, limestone blocks over-grown by vegetation, littered with shed snake skins and butterfly wings. Robbie and Bernadette warned of dangers all the way down: 'Don't touch that tree – it burns,' and 'Don't knock that spiny bush, there are fire ants living inside, and that'd be a trip to the hospital.' This, combined with some of the most abundant and persistent mosquitos imaginable, foreshadowed what we could expect when we properly had to go bush.

There was a choice of several cenotes, but we took the largest, with the cleanest entrance to the water. Gearing up in the mud and scrub was a different ball game to getting ready in a nice open car park as we'd been doing up till then. Suddenly nothing seemed to fit, my dive kit no longer hung right, every hose and wire seemed to take an eternity to slot in, and, at 100 per cent humidity, inside my neoprene wetsuit a generous puddle of sweat was gathering in my shreddies.

When I connected the main air hose to my mask, I noticed an unusual bulge near the connector. 'Hmmm,' Scott pondered, 'that doesn't look great.' As I made to respond that I was sure it would be fine, there was an explosion and the roar of 2,000 litres of com-pressed air rushing to be free. I was deaf in one ear, and totally disorientated, but Scott soon had the tank shut down. All the divers stood around looking quizzical.

'That was a catastrophic failure,' said Scott, 'I've never seen that happen before.' Neither had Robbie.

'That isn't supposed to be possible,' Bernadette said. This, from divers with tens of thousands of hours of experience between them.* Scott's face though said it all. If that had happened during the dive, it would have been really, really bad.

At that moment, James lightened the mood by sitting on a limestone block the size of a fridge and toppling it onto his foot.

Once we'd managed to get all my equipment and cave diving techniques focused, the rest of the known cenote diving was really pretty special. We dived alongside a small American crocodile, probably about the same length as me, dazzling as it swam over my head and straight at Robbie underwater, who ducked as if he was getting out of the way of a low-flying pigeon. Large ominous tarpon hung in the water. These fish, which are about the size of my leg,

---

* 'That isn't supposed to be possible' is a phrase I've heard many times with the Diveways. The Diveways is the mask that I use on all my dives. It's not like a regular scuba mask, but has an airspace over your whole face so I can talk to the crew while submerged. It's a great bit of kit, but is actually quite dangerous. As you are breathing out into the airspace in the mask, carbon dioxide builds up, and you then breathe it back in. If I'm talking for a few minutes, or swimming hard, I'll get a headache which tells me I'm taking a CO_2 hit and need to flush the mask out with air. Take too great a hit and you pass out, game over. Ten years ago when I started using the mask, I'd have to have a helper on the underwater communications system every few minutes saying, 'Purge, Stevie, purge,' as if passing on some thoughtful lifestyle or philosophical advice. Thankfully, I have now done well over a thousand dives on the mask, so have the CO_2 bit under control. However, on two occasions the Diveways has just stopped working. No rhyme or reason, and nothing we've been able to address or fix, it's just stopped giving me air completely, and I've had to rip the mask off and get to the surface. 'That isn't supposed to be possible,' is exactly what I've been told then!

look as if they've been wrapped in Bacofoil, and are actually marine predators. The fact that they're in the cenote is testament to the fact that it connects directly to the sea, via a continual thoroughfare of saltwater.

We penetrated numerous cave systems, swimming up and through the most remarkable haloclines, where denser saltwater sits beneath a 'river' of freshwater. The concept is relatively simple, but the actuality is one of the most bizarre distortions of perception achievable without psychedelic hallucinogens. Swimming into a cavern that Katy had carefully illuminated with a few strategically placed lights, it was as if a mirror had been placed below us, reflecting the stalactites trailing from the ceiling above. But it's as you drop down to the mirror that the real distortion takes over. Anything that touches the halocline seems to lose focus, as if seen through a lens half smeared with Vaseline. You drop below it, and the water appears clear once more, but then bounce back up, through a swirling greasy layer of stirred vodka Martini, and then you're flying, airborne in water as clear as air. It is one of the great wonders of the world, but one that only those who cave dive ever experience.

With the training and preparation complete, it was time for the expedition proper to take flight. Literally. A helicopter was the only way of us locating remote cenotes that could be the target for our diving. I had imagined hefty patches of forest, with a few tracks and trails, and maybe some small settlements. But instead, beyond the beaches it was just a blank expanse of green. Totally flat, featureless forest all the way to the horizon. Literally hundreds of miles. Standing out of the helicopter on the skids looking down with my binoculars into the forest was utterly exhilarating. It was true! Adventure and the future of exploration was alive and well in the Yucatan. This could be the most productive place on planet earth for exploration.

Two hours later, though, after circling the same patch of forest as slowly as possible, staring down through the foliage, my excitement had turned to concern. We were in an area where we had expected to see endless cenotes, but the forest was so thick we might as well have been trying to catch a glimpse of an ant family picnic on the forest floor. Nothing. The forest itself looked impenetrable, and it would clearly be a thrash to travel any distance at all. Plus we were talking about hacking through it with our scuba cylinders. My shoulder was still bruised from doing a few hundred metres. No wonder none of this stuff had ever been done! At our planning meeting that evening, the excitement was tinged with intimidation. This was much bigger than we were prepared for.

Despite these massive logistical concerns, it was something more mundane that brought us to a shuddering halt. The day of the halocline dive I'd had a nagging thud in my right ear. Ear infections are frighteningly common in diving, and I should have done something about it, but the film crew had only just turned up and I felt I ought to crack on. Big mistake. A couple of hours underwater and under pressure followed by a sleepless night, and the ear had inflamed to the level where I couldn't hear, and couldn't even chew my food.

The doctor clearly saw endless cases of dumb divers who wouldn't listen to their bodies, and was pretty matter of fact about the diagnosis and the treatment. Antibiotics, ear drops and . . . no diving. For at least six days. I left the hospital in a daze. That was most of the time we had set aside for exploratory dives. Just days into the first trip of my big year, and everything was going to grind to a halt. Six days later, I was still on three different courses of antibiotics, was suffering sleepless nights due to fever and I had a raging case of thrush in my throat. It seemed like my long-fantasized expedition was crumbling at the first hurdle. Truly the cenote gods had let us know how they felt.

While we were bound to dry land, we drove out into the forest to see if we could find some wildlife. In amongst the collapsed cenotes, we heard the loud calls of the motmots, orioles and the occasional ocellated turkey, and found skittering lizards, leafcutter ants trailing sails of flowers and leaves in their mandibles, and a fair few snakes.

Tucked into one particular rock crevice was a Mexican night snake, a snake I know all too well. In a cave not far from here, I'd spent five nights on my own crouched down in the darkness trying to film these snakes lashing out at and catching bats on the wing. Then, clambering over the limestone blocks, Mark, one of our cameramen, shouted out 'Snake!' He'd nearly put his hand on it! This was what local people call 'nauyaca', or the four-nose snake, because it appears to have four nostrils. Two of these, however, are heat-sensitive pits, which this viper uses to sense its prey. I squatted down with my face right next to it. Perhaps having it within striking distance of my nose had me distracted.

'Snakes like this are so well camouflaged, it would be easy to step on or near one and get bitten. They're the reason we wear high-heeled boots in the jungle.'

'I think you mean high-ankle boots, Steve,' Mark corrected.

We were here in the few weeks of the year when the 'flamboyan' trees were in flower, bursting forth with waterfalls of burning red blooms. We were also witness to the two months of the year when blue morpho butterflies are on the wing. This produces one of the most spectacular sights across Latin America. The butterflies' wings are a handspan across, and in the males the upper surface appears a flashing neon blue. It's only the males who bear these striking colours, with the females being much more dowdy brown, so the display is a sexual characteristic to woo the ladies. The blue is not due to pigment, but microscopic iridescent scales in tetrahedral shapes, which reflect the light in such a way that the wings appear to be illuminated from within. Like most butterflies, the eating and growing stage is the caterpillar; as adults they may use their watchspring-like proboscis to lap nectar, or to suck the fluids from dung or decaying bodies – even the salt-rich sweat from our expedition clothing when we hang it out on a line. The ancient Maya believed these dazzling wonders were the re-incarnations of warriors who had died in glorious battle. That glamour doesn't last long, though they live just a hundred days from egg to adult, and spend no more than a month as a flamboyant winged adult.

When the morning finally came for us to head out to the jungle in search of cenotes, my ear was still thudding a bit, but my blood was thumping in my veins so intensely that I barely noticed. As we were expecting there to be rope and clambering challenges, Aldo Kane had joined the team. An ex-Marine sniper and omnipresent on all my recent expeditions, Aldo had been nominated as safety expert, otherwise known as 'Head of Peril'. It was just like the *Avengers Assemble*. Except instead of Ironman we had Robbie. And instead of the Incredible Hulk, we had an angry bearded Scotsman.

The path into the forest started to the north of Tulum. To get out into the bush we first drove right through the city dump. A vision of post-apocalyptic hell, the dumps in huge cities like São Paulo Mexico City and New Delhi are the size of cities themselves, with communities of dirt-poor 'untouchables' picking a living in amongst the rolling, decaying stinking mountains of human waste. Tulum was a city of only 50,000 as opposed to Mexico City's 20 million, but its dump still covered undulating hillocks of blazing or smoking plastic, car tyres and fly-ridden food waste.

The carcinogenic fog that turns the sky above black is mirrored by black slime trickling down through the ground. And while elsewhere that will be filtered by soil and then solid rock, here in

Yucatan the water table is right below the surface. That freshwater is connected to all the rest of the freshwater here, flowing into the cenotes that Yucatan relies on for tourism, where thousands of people swim and dive every day. And of course to the watercourse that everyone relies on for drinking water. It was a sobering moment. Car Wash, where we did our very first practice dive session, was no more than a kilometre away, 'downstream' towards the sea. I was lucky to get away with just an ear infection.

An hour out of the dump, and we could still smell the acrid plastic stench. It would be stuck in our clothing till we left the forest. The grimness of it all was soon forgotten, though, in what almost instantly turned into a pretty committed hike. Robbie had got a hold of a few local porters to help us carry our loads towards the unknown cenote, but three of them had failed to show, so we had to apportion their loads between us. Aldo and I tried to outman each other by taking even bigger packs than the other, and we both ended up carrying around 40 kilograms, but badly loaded and with bits sticking out all over the place. I had my extra-long freediving fins crammed in to the side, and the drone bag clasped across my front. This was fine to begin with, but as the forest got thicker, every vine, bush and spine seemed to be deliberately trying to rip the bag off my back, or take me clean off my feet. I doubt two more sweaty people ever existed.

At one point we stopped beneath a stand of paper bark trees to witness a commotion in the canopy above – Geoffroy's spider monkeys, which were taking exception to our presence. They brachiated hand over hand into clear patches in the canopy to stare down at us and furiously shake the trees before weeing on us from a great height. All of this was a good sign, though; in forest where they encountered people often, they would have bid an instant retreat – even more so if they were hunted.

It was a surprisingly brutal trek, luckily only taking around four hours, because we all managed to drink our entire six-litre water allowance before we arrived, most of which had not passed 'Go', and had just pissed straight out of our skins. But eventually the buzz of the cicadas and occasional cackle of laughing falcon was overwhelmed by a familiar call. The onomatopoeic, repetitious cry of our motmot; the voice of Xibalba (the Mayan Underworld).

We approached the cenote with our guts in our gullets. Robbie led the way to the edge of the hole, with Aldo and Katy following close behind. My breath caught in my throat. The hole at the surface wasn't big, perhaps ten metres in diameter. And we could see now why it wasn't visible from the helicopter – the canopy closed completely over our heads. The trees that encircled the pit had long twisted tangled roots free-hanging down into the darkness.

I had carried with me a monster torch, in order to get a sense of what was below us. When I cast its brilliant beam down into the gloom, we all gasped collectively. The cavern was huge, a stupendous cathedral whose walls were beyond even the extent of my light. Disgruntled motmots raced in and out, perching on the dangling tree roots, perturbed by this unexpected intrusion. The water below us gleamed aquamarine and black; a film of calcite coated the surface, confounding any estimates of depth and scope. The otherworldly alien underworld had an energy somewhere between fantasy and nightmare – a preposterous *Pan's Labyrinth* distorting mirror of the real world. It looked like adventure.

We barely stopped to let the sweat dry before making a plan, charged with the thrill of what lay beneath. Aldo set up a simple rope system, while Katy, James and co. started getting cameras and kit ready for a dive first thing the following morning. That was never going to work. To be honest, it was all anyone could do to prevent Robbie and me just leaping over the edge with fingers crossed it was deep enough. Within an hour of arriving I was standing on the edge, in my bare feet, with just a mask and snorkel, preparing to rappel down into the gloom.

Suddenly, out of nowhere, I was hit by a sensation I hadn't felt for years – in fact, not since childhood. It was that deep-seated primal fear – though not of the drop below me, nor the potential dangers of the exploration. It was an unexpected and powerful chill, the fear of monsters under the bed, of trolls beneath the bridge, of witches in a midnight forest. I was frightened! I'm in my forties. I don't believe in ghosts. I've spent night after night on my own in total darkness in bat caves and never got the heebie-jeebies. But something about this place had me completely creeped out.

I muttered something trite about the privilege of being the first person ever to descend into the depths of the cenote, before shuffling back over the edge, and into another realm.

The overhangs beneath my toes were covered with paper wasp

nests, like Chinese lanterns or Christmas decorations, but with far more lethal portent. These can be one of the most deadly things about working on ropes in the tropics. Disturbing them en masse can lead to swarming attacks, resulting in hundreds of stings, an overload of vespid venom, and almost certain anaphylaxis, which out here would be fatal.

'Dozens of wasp nests under here, Aldo, old bean,' I said, trying to sound nonchalant.

'Water looks deep,' he replied, 'and it's only 15 metres down. If they go for you, I'll cut your rope.'

He wasn't joking.

I descended slowly, soaking in the atmosphere, but also delaying the moment when I would drop into the dark waters and be truly alone. I used my toes like a chimp to grasp the rope and slow my descent. Dropping below the level of the ceiling, the reflected light for the first time revealed the extent of the cavern. It was half a football pitch in length, and perhaps half that again in width. Giant stalactites hung from the ceiling, and the pool of light in dead centre was dominated by the suspended roots from five trees. These ended at the water in unusual bundles, tendrils sucking up the liquid and transporting it up to the trees way above. Something about their symmetry in the odd light made them seem like sentinels, or the columns framing an unusual sacrificial altar.

I couldn't hold off any longer, the ropes dropped into the water, and I had to as well, plunging into the cold blue, unable to see the bottom below me. Odd white wriggling shapes like thumb-sized leeches scattered beneath me, and then something grabbed my shoulder. I leapt and shouted in fear, swiping it away, my heart pounding.

'What was that?' Aldo's voice echoed down from above, reverberating around the tomb with the motmot calls and the flutter of bat wings. I grabbed my mask and pressed it to my face, squinting through the sweat and fog. Before me were two fist-sized, prehistoric-looking bugs, with brown carapaces like vast cockroaches, raptorial pincer forearms clasped in front of them. One had grabbed me.

'Water scorpions!' I sputtered. 'Bloody big ones.'*

I squinted further around the root balls, at the myriad white shapes skittering about in the tendrils.

'Bloody hell!'

'What is it, mate?' Aldo yelled down in concern.

'Blind cave fish,' I shouted back up, 'millions of them, and cave shrimp, and isopods. I've never seen anything like it.'

Cave-adapted wildlife in cenotes is neither diverse nor abundant. After all, it is a challenging environment where organisms exist in a fine niche, depending on organic material from the photic (light) zone, but protected in terminal darkness. Blind cave fish

---

* Water scorpions look superficially like scorpions, with a long slender tail (which is actually a breathing tube), and grasping limbs, but are true bugs, so related to assassin bugs and water boatmen. Records suggest the largest known species are only about six centimetres long. These, though, were huge horrors over double that size, and I know even little ones have a fierce bite.

that wander into the light are inevitably picked off by predatory freshwater fish. Because of this, they are rare sights, and it's unusual to see more than one at a dive site. Here, I could see hundreds.

'Are you going to let go of that rope?' Aldo shouted down.

'Yeah, alright, mate,' I shouted up.

'I can't send Robbie down till you're clear.'

'Give me a chance,' I yelled up, annoyed.

I pretended to myself that I was cross because I hadn't had time to take myself off the abseiling system and get myself sorted. But that wasn't it at all. In actual fact, I didn't want to let go of the rope, my one connection to the security of the team above. I didn't want to swim off into the darkness on my own. I was afraid.

'Robbie's coming in his pants,' Aldo shouted down. That broke the mood somewhat.

'You what?' I retorted.

'Well, it was that or his jeans.'

'I have another dive vest in my bag,' I shouted up, 'get him to wear that.'

I let go of the rope, and because I had no helmet on to protect me from anything falling from above, swam away from the roots and into uncertainty.

As I kicked off, the sun must have come out from behind a cloud, and everything changed. Shafts of brilliant light cut down through unseen holes in the ceiling – more brilliant than the limelight for a big solo show on Broadway. When the sunbeams hit the water, they cut down into the depths in turquoise columns. It was truly breathtaking. Just then, Robbie dropped over the edge and into this Hollywood film director's dream. In his Y-fronts. I swam over to him, and we clasped hands and laughed like drunken Wall Streeters who'd just sealed a big deal. This place was a natural wonder, and we were the first ever to see it.

Our initial recce was both exhilarating and intimidating. Directly below the collapsed ceiling, and beneath each of five 'skylights' in the roof above, was a mound, composed initially of the limestone from the collapse, and then from millennia of leaves, branches and other organic debris. In the centre, at about eight metres depth, was a spooky tree that may well have fallen in all the way back when the roof caved in. There were also several animal skeletons that had been picked clean by the isopods – scavenging white woodlice relatives the size of a grape that could pick a body down to the bone in hours. One of the skeletons was a perfect motmot, still entirely intact, and even bearing its tail wires with the feathered pendant at the tip. Its beak was wide open as if screaming.

Underwater, vast stalactites hung down into blackness, and beneath them the cave floor was invisible, but certainly far, far away. And through it all swam the blind cave fish, elongated shimmering stars in a night sky. These are cave brotulids, capped by an oversized ogre-like head, reputedly able to swallow prey larger than themselves. The body is elongate and eel-like, with a fin to the top and bottom that undulates to provide subtle movements, though occasionally they would accelerate away by twisting their entire body. It was the amount, though, that was really surprising. Somehow, here, in the absence of predators, they had exploded to incredible numbers.

We finally found a tiny sloping ledge we could teeter on, having been treading water or swimming for well over an hour. It was at the top of a vast stalactite the size of a column on the Parthenon, plunging down into the depths beneath us.

'I've never seen anything like it,' Robbie gasped. 'It's beautiful. And there are more blind cave fish here than you'd see in a lifetime.'

And when the greatest living cave diving explorer says something like that, you know it's special.

We hauled ourselves out of the cenote when Robbie was so cold he was literally blue and my teeth were chattering so much that the guys above could probably hear them. At the surface Robbie and I were bursting with enthusiasm, desperate to tell everyone what we'd seen. When we were both up, Robbie went for the handshake, and I went for the man hug, which was embarrassing. I styled it out, and grabbed a hold of him anyway, while he battled to try and make do with a fist bump. Awkward.

That night we sat talking nonsense around the campfire, charged with the thrill of experience and potential. Aldo told tales of his days in the Marines, Scott talked about shooting guns, riding Harleys and diving disasters, and Katy told us all about her rise from uni dropout to world-class cave diver. I just sat back and listened quietly, content, and actually rather moved by what we were here to do. Pretty soon the chat turned to bugs.

'These mosquitos are evil,' shuddered Katy, 'and I brushed an ant tree today and got bitten on the eyelid. They really hurt!'

The ant trees in Yucatan were spiky bushes with hollow thorns, and stinging ants lived within, protected, and providing protection for the tree, in a classic case of symbiosis.

'How come you never get bitten, Stevie?' James asked.

'I do,' I responded, 'I just don't react to them like you do. Just as well, really.'

James had been nailed repeatedly by horse flies, and had one hand so swollen it looked like an overinflated washing-up glove, and one upper eyelid was almost closed.

'I mean, you look like you've taken a right hook from Anthony Joshua,' I said.

'And like you've got a nasty case of elephantitis,' Aldo added.

'I definitely get bitten though,' I mused. 'I've picked four ticks off myself in the last hour, including one right on my danglies. I thought it was a big pus-y zit or something to begin with, but it turned out to be a nice fat tick.'

'Do they have Lyme disease here?' Aldo asked.

'Not sure,' I said, 'but they're bound to carry something unpleasant.'

Within a minute or so, everyone round the fire was subtly rolling up trouser legs and checking around their belt lines. Then everyone quietly got up from the fire and wandered out into the forest to carry out a more intimate examination. Nobody had escaped un-ticked.

Though the forest around us was painfully dry, closer to the cenote wildlife was abundant. As night fell, the motmots and cave swifts retreated into the caverns to sleep and the bats emerged, swirling around in the mouth of the cave as if getting up the motivation to head out and feed. Bugs were plentiful. The mosquitoes were also pretty robust, biting clean through clothing and hammocks. Small tarantulas scampered around camp – probably males looking for a mate – and scorpions appeared, one of which was a mother, carrying her tiny young on her back. As we approached, lighting her with a UV light, the babies all glowed bright yellow, like miniature glow-in-the-dark toys clustered beneath her curled tail. As in Western astrology and astronomy, the Maya recognize a constellation they call Scorpio, and it is the same as ours – a trail of celestial bodies trace the curled outline of a scorpion's tail.

Moths and big chafer beetles came thundering into our lights and caught on our clothes. Quacking calls from the mahogany tree frog ricocheted around the area immediately close to the cenote, but were silent no more than a minute's walk from the water. These tree frogs call out in the early evening to attract a female, then once they have mated the eggs are deposited into deep water. This is unusual – most forest frogs will drop their eggs into shallow puddles. However, in the Yucatan these don't exist, so dropping their eggs into deep water is the only option.

The next day was all about getting back down into our sacred chamber, with Katy on our one precious scuba cylinder of air to film, and also to take a closer look out to the walls and see if it turned into 'going' cave; that is cave that continues off in intricate passageways, potentially for many miles. Once inside, we knew we had a strict time limit. Katy only had a couple of hours of air to keep her underwater – less if she went deep, which would be extremely ill-advised as she didn't have a 'buddy' diver to rely on. We were also counting down the time, as neither Robbie nor I had carried in wetsuits, and so would only be able to stay in the water until we started getting hypothermic. Then we would climb up the lines, and within minutes be in danger of being hyperthermic (overheated)! We needed to use our time wisely.

To begin with, we focused on illuminating the extent of the cavern as much as we could, getting a sense of its limits. Katy's massive dive lights swept the gloom aside, backlighting the grand stalactites. The cave floor beneath the mega stalactite seemed to bottom out at around 30 metres, before sloping steeply up to the central mound. Carving round, she managed to light perhaps half of the cavern, before it got to the point where her air started to drop into the danger zone, and we had no choice but to dive down and motion for her to surface. She burst through the calcite crust on the top, bubbling with enthusiasm.

'It's so gorgeous,' she said, 'I could shoot down here for years.'

'How's your gas, Katy?' asked Robbie, which seemed like a presumptuous question.

'I'm at 60bar,' she replied, 'I could easy get another 20 minutes?'

'That's pushing it too close,' Robbie said. 'Anything goes wrong and we can't get you out.'

'And besides,' I said, 'your teeth are chattering, and your face has gone the colour of a Smurf.'

Despite wearing a dry suit and thermal layer, Katy was obviously on the brink of being proper chilly . . . in the tropics. She wasn't lying about the bioprene. There was no option. We needed to end the dive and get her out of there.

Twenty minutes later, both Robbie and Katy were out. Despite my considerable layers of blubber, my teeth were vibrating like I was riding a penny farthing down a cobbled street. But I didn't want to leave just yet. I floated on my back, looking up at the roof, wondering if I would ever see this place again. My cave. This special, glorious, ethereal place that I had been given the honour of christening.

At the surface, we sat around the smouldering campfire and had a council of war.

'You know you get to name it now?' Robbie said.

'Wow, really?' I enthused. 'What, like the Mayan word for motmot? Or perhaps we could just name it after the ancient Mayan for gateway to the Underworld?'

'I was thinking Backshall's Backdoor,' said Robbie.

The conversation then turned more seriously to what on earth our course of action could be.

'So if we want to dive this properly,' said Scott, 'it's going to be a big deal. Four divers in the water on twinsets or sidemounts;

we're talking eight cylinders minimum all the way through that forest.'

'Can't we cut a DZ?' Aldo asked, using the abbreviation for dropzone, 'and heli them in?'

'Or we just napalm the whole forest and be done with it,' Scott added laconically, 'then we could just drive all the way.'

'Yeah, that's you army boys' response to everything, Aldo,' I added. He was riled for a second, ready to correct me on the whole army/Marines thing, but then realized I was teasing him and cast his eyes skywards.

'OK,' James said, 'let's run through our realistic options. The heli is beyond our budget even if we did want to desecrate the jungle. Carrying eight tanks in through that trail would break us . . . we need to know it's going to be worthwhile.'

'I don't think it is,' Robbie chimed in. 'I mean, I know we only looked at half of it, but I don't believe there is going cave here. The geology doesn't fit. I think it's just one big chamber.'

'But it looked as if it went really deep over that side?' I pushed.

'Yes, so it could be like Angelita, dropping straight down to 70 or 100 metres,' Robbie countered. 'Do you really want to be doing technical diving that deep out here?'

'You'd be talking 20 tanks or more,' said Scott, 'and just the deco could be four or five hours.'*

---

* 'Deco' is nothing to do with painting and decorating, but is decompression – the time after deep dives you have to spend waiting in shallow water for

'Anything goes wrong and help is a very long way away,' Aldo added. 'I'm not carrying your heavy arse.'

'Specially not with all that bioprene,' added James.

'I could be wrong,' said Robbie, 'I just bet it doesn't go.'

'But you're the expert, and your opinion is everything here,' said James.

'Can we not just bring back one tank? Do another sweep of the other side?' I was getting desperate, and didn't want to give up my prize.

'My honest opinion,' said Robbie, 'it's not worth it. And I have another place I know will have cave. Somewhere we can dive properly right from the start. I promise you, that is a better option.'

Two days later, we were hiking off into the forest again. This time we were laden with even more weight – scuba equipment, and everything we needed to dive into the system Robbie merely called 'No Name'. He had identified several different potential entrances, but had never had the chance to come back and dive them. They were not cenotes like Backshall's Backdoor, but small caves with unassuming little pools inside. This, though, didn't mean anything. Most of Robbie's biggest cave systems had started just like this, tiny inconspicuous entrances leading to miles and miles of staggering sunken cave.

After a couple of hours walking, we found ourselves coming

---

nitrogen bubbles to dissipate from your blood. Spending five hours hanging around underwater is not my idea of a good time.

down to a cave entrance perhaps waist-height and four or five metres wide. Once we ducked down inside we found a pool that was actually more like a puddle. When we put our hands into the water tiny little fish arrived and started nibbling dead skin off our fingers. These 'pedicure fish' were apparently quite a good sign, showing that the water is being continually replenished and possibly linked to other larger water sources.

In my overenthusiasm to prepare for the dive, I put my wetsuit on as soon as we arrived, and once I'd carried a few scuba cylinders down into the cave, I started to turn into a boil-in-the-bag supper, beetroot red in the face and dripping like a leaky faucet. As there was clearly a while to go until we were ready to dive, I took myself down into the cave and waded into the water. I'd stepped up the thickness of my wetsuit and unwittingly made myself super buoyant – so much so that I floated at the surface exactly as if I were in one of those float tanks that were popular in the 1990s. In the darkness it was preposterously relaxing, and as I floated there with the ceiling mere inches away, my mind started to drift, and before I knew it the last few days of effort washed over me and I fell asleep. Minutes later, I was woken by a laser light bite to my earlobe. I screamed in reflex reaction, jumped up, bumped my head on the ceiling and swore loudly.

'What? What is it?!' James shouted in concern.

'Nothing,' I replied sheepishly, 'one of the pedicure fish just bit my ear.'

As if it wasn't enough to be attempting genuine exploration on camera, we decided to use the Diveways mask. We'd deliberated long and hard about this. Wearing this mask meant I wouldn't be able to wear a helmet, which is normally essential for exploration. What's more, we had to think incredibly carefully about our safety systems. I could not be in any way reliant on the other divers. If anything went wrong, I had to be able to get out of the cave under my own steam and with my own air. Otherwise I would be risking the lives of my entire team.

Our challenge was to ensure I had two independent systems and could bail out to either of my side-mounted cylinders, taking the full-face mask out of the equation. We decided to use a caving sidemount system, but with extra regulators on each cylinder, so if anything went wrong, I could rip the full face mask off and transfer to a normal mask, but critically still breathing off my own air and still able to obey the rule of thirds. I used a switching block on the mask to move back and forth from the two cylinders, otherwise one would decrease in pressure faster.

This was the moment when everything got real. Once we were underwater, we were on our own. There would be nothing our team at the surface could do to help. As Aldo commented, 'They might as well be in space.'

Scott, Katy, Robbie and I gathered in the tiny cave, gearing up and making sure every last thing was checked, double- and triple-checked. Just as we were about to duck under, I noticed one of my hoses into the mask was fizzing and surfaced. It was happening again! I shouted to Scott to come and help me, showing the escaping air.

Silence. You could feel the sick despair from the whole team. All this way, all this effort and the damn mask was about to destroy all our plans yet again. Scott's face was grim, but he whipped out a multi-tool and started to tinker. A couple of minutes of screwing and tightening later, and I turned the air back on. No more fizzing.

'Should be OK . . .' Scott said, hesitantly. Now that's the level of confidence I needed before plunging into a life-or-death dive.

'OK, I go first,' Robbie said. 'And I lay the line.' The line is a reel of safety cord that divers follow, which is also known as the Hansel and Gretel escape route. As long as it wasn't eaten; those pedicure fish looked pretty hungry.

'Steve follows next,' Robbie continued. 'Be very careful, do not kick up the silt. If viz goes, we have big problems.'

With that, he breathed out and slipped below the surface, the lights on his helmet immediately casting a ghostly rippling light over the ceiling.

I gave him a few minutes to secure the line and followed. The last thing I saw through the swirling brown waters was the filming lights of my crew. I couldn't see their faces in the darkness, but knew they'd be creased with concern that wouldn't cease until we were back safely.

Below the surface all was brown fog. I put my hand to my mask to wipe it clean. I couldn't see my fingers. Groping off into the swirl, my fingertips grazed the line and I took hold of it as if my life depended on it. In my cave training I'd done countless drills in total darkness to simulate these conditions, so knew not to panic. I had enough buoyancy in my system not to lose sense of where was up and down. I just needed to descend slowly and easily, doing my best not to disturb the visibility any further. These caves were legendary for the clarity of their water, so it would clear, I just needed to get down below the sediment. Clearing my ears, I descended carefully, slowly and easily. Nothing. Still brown soup. OK, now my heart was starting to beat faster, and my breathing rate was increasing. (Keep it together, Backshall, keep it together.)

Ten metres in depth, twelve. I bumped along the bottom, with one hand over my head to prevent me whacking it on a rock ceiling I couldn't see. I gently eased myself down the line, not moving my fins a millimetre. It made no difference. This was getting serious now. We were in a narrow choke, and 16 metres down. If something went wrong it would be highly unlikely I'd get to the surface. At that second someone bashed into me. All I saw was blinding light, dazzling, disorientating. Then Katy's eyes came into focus. Too close! She was supposed to be way behind me. I must have been moving really slow.

'Can you hear me, Katy?' I spoke into the microphone. She nodded her assent.

'What the hell's going on? I thought we would be out of this by now.'

She shrugged, an exaggerated gesture to show she didn't know.

Then white eyes swung into focus. More blinding light in the eyes. It was Robbie. He gave me a thumbs-up right in my face.

Really? It didn't feel that great to me! Then I remembered our signalling. A thumbs-up signalled 'up'; an instant return to the surface. We were bailing! My heart thudded.

'Katy,' I shouted, as if trying to actually shout through the water, 'we're pulling the dive, head for the surface.'

The brown fog swirled in front of me, and a fin came out of nowhere and twatted me on the side of the head. A mask strap pinged off, and air started frothing out the side. The bubbles raged off the low ceiling, dislodging loosely crusted calcite, which rained down over my head. Things were going from bad to worse. With one hand still clasping the line, I reached up to clasp my mask onto my face, and blasted my own 10,000 lumen dive light into my eyes, frying my retinas like huevos rancheros. Disorientated, frightened, mind racing, I thrashed upwards, and broke the surface just before Robbie.

'Well, that was interesting,' he said, with generous understatement. 'We kicked up too much silt getting ready up here, so we wait ten minutes, and then I go down, find the way on, and you all follow the line.'

That was when I loved Robbie the most. This guy is the real deal. He was trusting enough to take me along on exploration of caves he'd sweated his guts out to find. Then I turned up, destroyed the visibility and made a circus of the whole thing. He could easily have balled me out for being hopeless. Or stormed off and refused to work with such amateurs. Instead, he took a deep breath and tried again. For ten minutes I sat picking chunks of chalky rock out of my scalp (I'd have dandruff made of 100-year-old calcite for the next fortnight) while we waited. Ten minutes later, we tried again.

The entranceway tube was still cloudy, but I moved down deliberately and calmly, hand over hand down the thin string of the guideline. And then the soup became clouds, and the clouds started to billow and then clear. It was like drawing back the curtain to reveal another world. The first cavern wasn't big, nor particularly ornate, but the water was crystal clear, and Robbie flew ahead of me, frog kicking effortlessly. Shining ahead of him, his light illuminated the cave in stunning shades of blue. Aquamarine in close, through tones of velvet royal blue until, beyond the range of the light, it faded to inky black. Robbie's exploration was remarkably quick; his understanding of where the cave would go instinctive and decisive. When we came to a big room he'd pitch around it, searching for exits; when we came to a junction he'd signal me to hold, before recce-ing all the different options and returning with the best choice.

About 15 minutes in, and the tunnels opened out into a larger room. The chamber was revealed behind a veil of stalactites so dramatic they looked like the pipes of the Royal Albert Hall's organ. The ceiling was adorned with straws of calcite so brittle and fragile that the bubbles from a single exhale would have brought them down in a calcite cascade. The grandeur of the room, scythed through with dive-light laser beams, was beyond belief. And we were without question the first human beings ever to see it.

Around half an hour into the cave, it narrowed into a

constriction. Thus far, I'd been doing all I could to keep my cylinders from making any contact with the bottom and fins from even creating a wake that might dislodge something precious. Now that went out the window. The only way to progress was scraping along the bottom, with mask rasping across the top of the ceiling. And it wasn't short either – we swam for at least 30 metres or so through the crack, which probably marked the weakness at a bedding plane.

Halfway through, Robbie gave me the 'hold' sign again, and I lay on the bottom and waited. It was the first time I wasn't thinking about all my safety checks, and stopped distracting myself with not destroying all the stalactites like a swimming bull in a subterranean china shop. And that's when it hit me. I was 300 metres into an underwater cave, in a squeeze you couldn't get a beach ball through. That's when the weight of the rock above seemed to come bearing down on me. I needed to get out, to get to the surface, to be anywhere else but here. I needed to take my mind away. So I reached for my air gauges. I turned the right one towards me, but it was dark and I couldn't see the dial. I shone my massive dive light towards it, but it was too bright, and flashed off the dial.

Starting to get thoroughly panicky again, I reached for the other gauge. But there was a big patch of condensation in the right corner of my mask, and I couldn't see it at all. I flushed the mask out with an overlong burst of purging air, getting rid of all the CO_2, clearing the screen. Then I rolled to one side, pulled my cylinder into a good position, and bounced the light off the wall. I then did the same on the other side. So now I had my air values. And absolutely no idea what to do with them.

We started the dive with 3,000 PSI of air. We were supposed to breathe 1,000, come back when at 2,000, and leave 1,000 for safety. But in the aborted start I'd breathed rather a lot of air already, and so we'd recalculated. I'd started on 2,500. Or was it 2,300? Or was I supposed to come back on 2,300? Or was I meant to breathe 2,300? And my other gauge was totally different . . . what did I do?!

This is actually very common in diving. There's so much going on, and things like nitrogen narcosis can totally screw your brain. Even now, after 30 years of diving, if I'm opening or closing a valve on a dive, I have to recite in my head, 'righty tighty, lefty loosy', or I'll mess it up. In one of my training courses, we went into a compression chamber and were 'taken down' to 40 metres of pressure. We then had to do simple motor skills and word games. Most people could barely remember their own names.

And this was happening to me now. I tried over and again to work it out, but I'm not great with maths at the best of times. There wasn't really any choice: we were going to have to turn back. I could see Robbie's light coming back to me. He wagged his finger – 'That way's no good' – then twirled one finger around – 'Let's turn back.' He took the decision out of my hands.

Half an hour later we were back at the surface. James and Aldo had been starting to get decidedly anxious; we'd told them we'd be at most an hour and had been 70 minutes. But they were also elated to see the look on my face, and quickly realized that we'd seen something pretty special. When Robbie surfaced, we could instantly see he was pretty happy too. I lunged forward to give him a hearty handshake. He went for the fist bump. I ended up shaking his fist.

It wasn't until later, sitting with a beer watching Katy's footage, that I fully began to appreciate what we had seen and what we had achieved. We had chosen one of thousands of possible cave entrances in one small corner of Yucatan. Our team had swum in, and in one short dive uncovered half a kilometre of pure wonder. Real, proper, barnstorming exploration, no more than a metaphorical stone's throw from the biggest tourist site on the continent. Our project had been launched in spectacular style, and the Mexico expedition was not even halfway done.

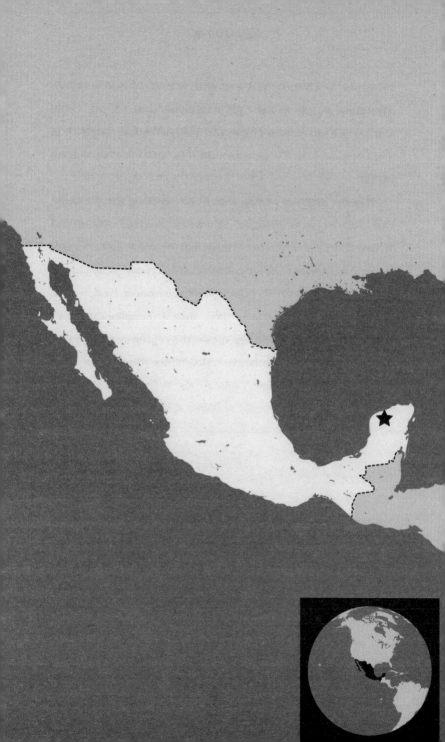

2. ANCIENT MAYA

MEXICO

Each of us steam with clouds of vapour, bringing to mind prop forwards mid-game on a freezing February morning. Sweat leaks down our faces and puddles in our boots. Ten hours underground, and each of us is drawn and haggard, hemmed in by the rock and the stale air, longing for a breeze and release from this dank sweatbox. Terracotta pots, carvings and shards of broken containers litter the cave floor. Charred shapes daubed on the walls depict leaping deer and snarling jaguar – sacred beasts with the power to enter the underworld of Xibalba, the place of fear. The soundscape of the cave messes with the senses: the limestone ceiling deadens our voices, making our conversations sound like tape recordings stripped of life. Yet a distant drip resounds with a generous ploppy echo, and the flutter of a bat's wings seems deafening. We instinctively whisper, as if tourists in a cathedral. Aldo drops a carabiner to the floor, and we all stop dead and wait for trolls to waken from the shadows. His eyes are surrounded with black rings of mud-run sweat, and his beard drips. Before us, a tapir appears to leap from the wall. It is a miracle of sculpture, perfectly moulded and shaded with charred campfire ash. All the more miraculous because it was emblazoned on the rock several thousand years ago, by a barefoot Mayan holy man, his work lit only by a flaming firebrand.

The halfway point of my six weeks in Mexico's Yucatan province marked a turning point in our expedition. From the coast, we were to head inland to the interior of the state, and the heartland of the ancient Maya. We also took on a new member of the team. This last addition was the most esteemed. Guillermo de Anda is a university professor, *National Geographic* explorer-in-residence, and the most renowned archaeologist in Yucatan. Having watched him in dozens of films, TED Talks and lectures, I held him in absolute awe. His discoveries and theories were legend, so I was a little nervous; meeting someone of his stature I always feel a little like a schoolboy being sent to the Headmaster's office.

In preparation, I had poured through books and countless articles on Mayan culture and its historical significance to the peninsula. I needn't have worried. On arrival, he greeted me like an old friend, made reference to a bunch of my own films, and sought to put me at ease and legitimize my position on the team by saying over and over: 'I'm so looking forward to getting your zoologist's opinion on x,' or 'It'll be so important to get your expert view on y.' (I'm sure he knew full well how clueless I was, but it was incredibly kind all the same.)

It was difficult to put an age to Guillermo – or Memo, as everyone called him. He was bald-headed but with a perfect dark goatee, smiley black eyes, and a posture and barrel chest that suggested he was once quite the athlete. In the weeks prior to joining us, he had been in a car crash and was lucky to escape without injury, and then had an accident with a camp stove and completely burnt off his eyebrows and much of his beard. It seems even the greatest of explorers can be a liability every now and again. Around his neck was a chain with a tiny silver old-fashioned deep-diver's helmet. His stories, enthusiasm and knowledge made us want to throw down our forks and just head straight out and start exploring!

While some academics can be a bit uptight, Memo was the opposite. Every dumb thing I suggested, Memo acted as if I had just discovered a brand new way an age-old mystery could be solved. When I mentioned something I had found in my trite research, he nodded his head sagely as if I were an Oxford emeritus professor. My kind of guy.

But it was Guillermo's tales that had me really inspired – more than I had been for many years. He explained that the entire peninsula was peppered with passageways and full of forgotten art and artefacts – the entirety of Mayan mystery frozen in time beneath our very feet. He spoke of an underworld where history came to blazing light, a place where we would rekindle the excitement of Howard Carter opening Tutankhamun's tomb . . . on every day of the expedition.

As a sort of intro to Guillermo and his work and passions, he'd arranged to take us to a cave that was easy for us to access. After walking no more than a kilometre down a dirt track, we plunged into the forest for a few minutes and found a small depressed entrance. As we stepped down inside, we became aware that we were on a staircase made out of the limestone blocks from the roof of the cave. La Cueva de Las Manitas – 'the cave of the little hands' – was remarkable for many reasons. Number one, it was not signposted and was hardly visited – in fact, it remained totally unknown except to a few archaeologists. And it was right by the road, 20 minutes from the nearest small town. In Europe, people would cross the continent to visit something so dramatic. Yet in Yucatan, with its world-famous sites of Chichen Itza, Uxmal, Calakmul and Tulum, it is a mere trifle.

I had deliberately not asked any questions about the cave and what it contained, but with something that was rather easier to get into than Wookey Hole (one of Britain's most famous show caves), my expectations were not high. We kept our lights off till we were well into the main chamber, and Guillermo gave the command: 'OK, Steve, fire 'em up!'

As we illuminated the wall and ceiling before us, we were rewarded with a remarkable sight. Probably 50 or 60 handprints, formed by blowing pigment over the top of the hand and wrist as a sort of stencil. Most of the pigments had been made from the red juice of crushed cochineal beetles, mixed with clay. The effect over deep time was that of dried blood.

A little further down the cave were reverse prints, where small hands and feet had clearly been dipped in the pigment and then pressed to the walls and roof.

'I've seen prints like this before,' I whispered, 'all over the world – Borneo, New Guinea, Europe, there must be something universal about it.'

'Yes,' Guillermo responded, as if he had never noticed this before. 'Exactly. And I think the two kinds of print are different.'

'The pressed handprints, they're kind of like a fingerprint, right? Leaving your own unique mark behind?'

'That could be right,' Guillermo agreed, 'but these small ones are from children, and the ones on the roof are feet – they did not press them there themselves.'

'Well, maybe the parents did it?' I asked. 'Kind of a record of them as babies, isn't it?'

'We don't think so,' Guillermo said. 'In fact, it's sad, but we believe the children whose prints were put here were later sacrificed – they were killed, then their bodies were thrown into the cenote.'

This didn't tally with my own culture's neat perception of how such things worked. Why sacrifice children? They are the most precious members of a population; why would you sacrifice them to the gods?

'It's complicated,' Guillermo said. (As a Latin American, he enunciated the 'cay' in the middle.) 'Firstly, you have to remember that death is seen as an honour. It is not a bad thing to die, and to die as a means to please the gods, for the Maya that is a privilege.'

That I could understand, that's a common thing in many cultures; certainly Islam, animist beliefs, and even Judeo-Christianity have the concept of heaven and 'going to a better place'.

'But then,' Memo continued, 'we found in the Sacred Cenote, 83 per cent of the sacrificed bodies were children, and almost all of them show signs of sickness. This was probably not throwing away their best people.'

He went on to explain that the positive and negative prints, though similar, may have diametrically opposed symbolism.

'Duality is everything in Mayan culture,' he continued. 'Pyramids are trying to reach the sky from the Underworld, up is down, down is up, often they are built on top of a cavern, on top of a cenote so the offerings to the gods can be thrown inside.

'So we believe that the two kinds of handprints are different. The positive print represents yourself – as you said, it is your mark, your *self*. The negative print could be the opposite . . . your alter ego, your spirit body.'

We took one last look around the cavern before heading on. Not for the first time, I felt a tickle of energy up my spine. This place was a cosmic spark plug, a zap of static connecting us back to the ancient Maya. More than a thousand years ago, a father had stood here with his child, wiped their hands and feet in pigment and pressed them to the ceiling. Then they had taken the child to a sacred sacrificial altar, and, using a knife of obsidian or a stingray spine, they had cut open the child's chest, and removed its still-beating heart, before throwing the body into a cenote. This as an offering to Chaac, the rain god, so that the waters would rise, and the people would not perish. Perhaps I only saw this moment so clearly because fatherhood was so nearly upon me. For a millisecond, that spark connected me and my Mayan counterpart, standing here in this very cave. I was genuinely moved by the experience. Even more so, because the very next day we would be diving in the very cenote that Memo believed to be the last resting place of those same sacrificed children.

The next day we pulled up at an abandoned hacienda, overgrown and crumbling. Haciendas were constructed by the emissaries of the Spanish empire, and were estates or plantations. Today, some are kept in fine colonial grandeur as hotels and restaurants, and some as private residences for drug barons and mega-wealthy plantation owners. Others, like this one, had faded as slavery in the region had become less formalized, and eventually collapsed into ruin. Everything about this place needed to be kept secret, for fear that bounty hunters would come and raid the cenote of its treasures; unthinkable, as the cavern was the sunken resting place of sacrificed children

Not surprisingly, the hacienda reeked of atmosphere. Some of the faded bright colours of the pink plaster still clung to the crude limestone blocks that made up the walls. The windows and roof design were deliberately flamboyant. There was a row of windows that were round like the portholes on a ship and a crumbling wall was topped with pointed castle turrets, but the grandeur had been no protector against time.

One courtyard held a rusting metal grate over a slightly elevated well shaft. As we pulled the grate back and looked down, we could see the transition of time. At the surface, a plastic pipe carried water away for a nearby modern farmer to water his cattle. Further down, this was plugged into a rusty metal pipe that looked a hundred years old. As my eyes followed the pipe down, the neat symmetry of the well became less and less defined, until it dropped into darkness. Here were two separate holes, and though they had clearly been enlarged, they probably represented the natural hole in the cenote roof. Beyond these holes, down into the dark you could just make out the shimmer of water.

Getting into the shaft and down to the water beyond was our first challenge. Aldo had the idea of using a frame that looked like it was made of scaffolding poles, and lashing it to the pickup trucks, giving a central point for lowering us down. We battled away for most of the morning getting it positioned so that it would offer a descent with the minimum of contact with the walls. There was no shade around the well shaft, and we all cooked in the midday swelter.

As soon as Katy was in the water, I needed to be on the ropes and ready to descend, quickly followed by Guillermo, so I got ready, wetsuit on, prepared to leap on the ropes as soon as my turn arrived. For one reason or another, that moment got postponed and put off even further, and pretty soon six millimetres of neoprene started to do its work. When I heard Katy splash down, I struggled into my cylinders and hauled myself up onto the ledge, clipped into the ropes, poised, ready, dripping with sweat. And waited, and waited . . . and waited. It was another half an hour, teetering on the edge, carrying a full person's weight over my shoulders in the form on my scuba set-up. By this stage, I was starting to superheat. Sweat was pouring out of my sleeves like I had a hose tucked in my armpits. My head was thumping, my heart was doing overtime, everything was starting to go wobbly. I was going hyperthermic – into full-on heatstroke. I've been through that before, and it's not pretty. I was going to have to pull out, to take a couple of hours to cool myself down. How stupid to be broken by something so mundane.

'How you doing, dude?' Aldo asked.

'Borderline,' I responded, making eye contact. He knew me well enough by now to know that meant trouble.

'Katy!' he yelled down, 'we need to get Steve down now!'

'I haven't placed my lights!' she yelled up, a far-off echoey voice.

'It's now or never,' I yelled down.

Silence.

'OK, I'm coming!'

I was carrying too much weight to lower myself down the well, and needed to be lowered by the whole of the rest of the team. It's an unpleasant feeling for any climber to be under someone else's control; you suddenly feel immensely exposed and vulnerable. I bounced my tanks off the rock around the well shaft entrance, and then was hanging in space in the darkness.

By now, the heat situation was too serious for me to take in my surroundings. I just needed to get into the water as quickly as possible. On splashdown, I instantly plunged my head under, seeping as much water as I could beneath the wetsuit rubber. It practically sizzled. My mask fogged straight up with all the humidity coming off my beetroot face. I floated on my back and paddled away from the light coming down into the cavern.

As I got out to the rock walls, finally I could take in my surroundings. It was midday, and the skylight in the roof held a vertical shaft of light so defined that it looked like Scotty was trying to beam someone up from the darkness. Without the *Starship Enterprise* light beam, the cave would have been invisible to human eyes. It was a big room, perhaps half a football field in dimensions, with a roof 20 metres high at its apex. Below the water, the light shaft was even more defined, an aquamarine dream shining down onto a central spoil-pile hill, which came to about seven metres from the water's surface. It looked as if there should have been a sunken treasure chest sitting on top of the hill, overflowing with pirate gold.

Now it was Guillermo's turn to be lowered down into the abyss. As he came over the brink he must have smashed the wall with a tank. There was a short call for attention, and then a block of rock the size of a toaster plummeted down and smashed into the black waters. Good job I'd swum away from the hole. From that height, even if I'd been wearing a helmet it would have staved in my skull like a cantaloupe.

With Guillermo down, it was time to dive. Thankfully it had taken half an hour for him to descend and by now my temperature was approaching normal, though sweat was still dribbling down my forehead inside the mask, even underwater. As we moved down the hill at the centre of the cave, it was instantly obvious that it wasn't entirely made up of rockfall. There were chunks of ceramic of uncertain age, bits of metalwork that had probably had connections to its function as a well, and bones of various animals. One in particular really caught the attention. It was the skull of a magnificent steer, bleached white, looking like a prop from a spaghetti Western.

As we continued down the hill and off into the cave, the creep factor increased dramatically. Firstly, more animal bones – dogs, cats and a dark brown horse skull. Then pottery – whole pots intricately patterned that could have been 3,000 years old. It was deeply odd. But that was just the beginning. No more than 20 minutes into the dive, we saw our first human remains.

'This is a femur from an adult male,' Memo said into his microphone, 'and that's his skull there. It's difficult to tell, but this was probably thrown in from the window above.'

'I don't understand,' I said, 'why they'd sacrifice people into their water supply. Surely it would taint the water they rely on?'

'I thought that too,' he said, 'but the truth is that the isopods and shrimps would strip a body to the bone in a couple of days.'

We swam out from the hill towards the walls, our torches cutting like light sabers through the black. Everywhere you turned your light there was a pot, a chunk of ceramic and huge amounts of bones. Some animals, many not.

In one position, there was an overhang of rock, and beneath it was an adult human skeleton, laid out perfectly as if it had been given a burial here. I asked Memo how it could have gotten here.

'It's hard to know,' he replied, 'but I think this was put here when water levels were way lower. Then maybe she was a sacrifice to the rain gods, an offering to bring back the rains.'

'Ten metres lower?' I asked, incredulous.

'Maybe. Or it is even possible they could have freedived it down here? We have evidence they did freedive into the cenotes.'

I imagined ancient Maya dropping down a vine or some kind of natural fibre rope, before swimming down into the darkness . . . with a dead body! Everything about this place was madness. The skull's empty sockets stared out at me, its lower jaw hanging low as if it were screaming. Or laughing at me.

Eventually we had all sucked our tanks dry and had to surface. We broke through the water, bubbling with adrenalin and excitement. The hour we'd spent submerged, paddling through a silent sunken graveyard, had been like swimming inside a beautiful nightmare.

That evening, we were many hours from any conventional place to sleep, so Scott spoke to the pastor at the local convent, who agreed to let us stay. It was a magnificent piece of conquistador-era architecture, with grand arches, red stucco plaster on the outside walls and cracked white plaster throughout the interior. We arrived as the sun was setting, and ran up the stairs that led to the roof. A mass was taking place in the chapel below, with hymns sung in Latin in a discordant minor key. On the roof, we stood alongside the three bell towers. The bleeding sun stained the town below us deep red.

Kids played football on the convent lawn, cockerels sounded in unison. Beyond our towering throne high up on the roof, the forest stretched off in every direction to the horizon, so flat and uniform it was almost unearthly.

It struck me that this was the first time since arriving in Yucatan that I'd seen a view. You are forever at sea level here, in a landscape that has no high points. No wonder then that elevation was so sacred to the ancient Maya, and they took such effort to build their pyramids that erupted from the forest and gave them domain over the landscape. For the kings, the high priests and the commanders of the armed forces, the benefit of height would be priceless. So it is also not surprising that when the conquistadors arrived, they usurped these sacred sites and built their churches, their convents, their monasteries on top. Thus, they would sequester their power and energy, and be a step ahead when it came to converting the heathens.

The conquistadors brought this idea with them from home, with the same process being common throughout the Christian holy lands. Most Christian sacred sites were not decided until 400 years after the supposed time of Christ, and then constructed deliberately on top of pagan places of worship. The Church of the Nativity in Bethlehem was built on the temple of the pagan god of agriculture, and the Church of the Holy Sepulchre (said to be where Christ died and was buried) over the Temple of Venus – goddess of sex and fertility. Neither site had been identified as Christian sites before the fourth century. This was a powerful way of taking on converts to the newer religion, and can equally be seen in our festivals: Christmas being conveniently situated over the pagan winter solstice festival of Yule (with its holly, ivy, Yule logs, presents, feasting and Santa-like green man), and Easter over the feast of the spring goddess Eostre (a vernal equinox festival of renewal with its symbols of bunnies and eggs).

We slung our hammocks up around the main courtyard, a rich orange glow painting the plaster as we told tall tales and joked about the dives ahead. Hanging above my hammock was a huge crucifix with an almost life-sized spooky Jesus, with lifelike blood and wounds on the yellowing enamel of the flesh. Not for the first time I pondered the absurdity of icons of a blond, fair-skinned Middle Eastern Jew, worshipped by both dark-skinned, black-haired Spanish colonists and the Amerindians they were converting.

Despite the grandeur and atmosphere of our camp, it was a sleepless night. And not because every time I rolled over I came face to face with creepy Jesus, seeming to condemn me for my sins. Instead it was down to a cacophony of snoring the like of which I have rarely heard before. Many insects such as cicadas practise a form of noise production called 'phase synchronization', where one wailing banshee sound starts to sync up with other cicadas, until the screams are constant. Several members of my team managed to do exactly this across all four quarters of the courtyard, with Stephenson's Rocket roaring from one corner, a revving-up Harley with a hole in the exhaust from another, and what sounded like a wounded bison adding its own flavour to the racket. Earplugs, pillow wrapped round head, and even a sleeping tablet did nothing to drown out the snoring. The next morning everyone was accusing everybody else, refusing to believe they were the cause of the kerfuffle . . . but producer Anna had gone around in the night with her phone and recorded the people who were most at fault. There was no escape!

Aldo and I were not on the list, but this was at least partly because neither of us had slept, which was unfortunate, because this was to be one of the hardest days of the expedition for us. Memo was taking us to a highly secret dry cave that was near to the temple complex of Calakmul, and had only been visited by a handful of scientists and archaeologists. It had been really tricky getting information on how we would access the cave, but we knew we needed ropes and climbing equipment to get back out. We over-packed, ready for any eventuality.

It was several hours of driving through dusty Mayan villages and thick, dry forests before we climbed up to an unusual spot of natural elevation, which must already have afforded the ground ritual significance. The cave entrance was hidden in amongst dense vegetation, in a narrow sandy-floored alcove, guarded by two big stalactite pillars as if the entranceway to another dimension.

To get into the cave we dropped down a three-metre step and into a little cauldron. We then stooped below a low ceiling, twisted and turned a bit, then fell down into darkness. After a few metres we came to the top of our first drop. The cave was hotter than a Turkish bath. We were drenched in sweat, steam pouring off us. At the bottom of this drop, we scrambled on down a tunnel with a low ceiling before arriving at the first proper drop, an overhang of maybe 20 metres down into a cathedral cavern that was truly vast, certainly bigger than the extent of our high-powered torches.

The cavern itself turned out to have abnormalities, which although an accident of geology, made it a remarkably important place to the ancient Maya. Despite the fact that we had needed modern climbing equipment and lights to get this far, and it had taken us at least three hours in total darkness, the central cavern was the gateway to a world as magnificent as anything ancient Egypt had to offer. The cavern had a big hill at its centre, and on top of it was a pile of stalactites, some of which had been pulled together into a crude altar. In the centre of that were a few pieces of smashed ceramic, and a stalagmite in an unmistakeably phallic form. It looked like something you'd get from an Ann Summers' catalogue, but in rock.

'In Mayan culture,' Memo told us, 'the stalactite is the feminine, the breast. The drips from the tip is mother's milk. The stalagmite is the masculine, the symbol of fertility. When the two grow together, there is procreation, the earth is born.' (Memo is the greatest tour guide in human history.)

'And this part of the cave, imagine how crazy this must have been for the first ancient Maya who came here. The entrance is on the highest point for miles around – sacred. It has the mountain in the middle – sacred; it has tunnels off north, south, east and west, at every point of the compass – sacred. It's no surprise they put so much effort into exploring here.'

'But how the hell did they get down here?' I asked.

'Who knows?' he responded. 'I used to think they might have built a tower to get down, but there is no sign of it, and wood lasts forever down here.'

'Ropes?' I suggested.

'Maybe,' he nodded. 'They had rope from material like hemp, and we know they used it to get into caves. But then how did they climb out?' He motioned to the drop in front of us. 'I mean, you're a climber, would you do that?' I looked at the slick, greasy, overhanging, featureless rock. No modern professional rock jock climber could get up that, let alone with a burning torch in their mouth.

'It is a mystery,' said Memo, 'but like so many of these caves, the challenge is the point. Perhaps even to get in here is a rite of passage, a test of fortitude. I have found caves near here where they have made artificial squeezes you have to get through, pits full of spikes you have to cross . . .' He smiled and shrugged. I held back the urge to mention Indiana Jones yet again.

Of the four tunnels that led on from the cave, Memo pointed us towards the northern one. It was a careful scramble through towering formations, and some of the most beautiful cave creations you will ever see. Those paled into insignificance, though, as we found our way into another giant cavern. Memo motioned towards black smears on one flat wall. As we turned our torches onto them, they transformed before our eyes, almost seeming to come alive.

'It's a jaguar!' I gasped. The mural of the big cat was

anatomically perfect, tail curled up behind it, looking back over its shoulder. It even had spots.

'And it's hunting,' I exclaimed. There was another animal in front of it. 'What is that, a deer?'

'I think so,' Memo said, 'but I don't think this is a hunting scene. Look how they are both looking over their shoulder – the jaguar is not looking at the deer, they are both looking in the same direction. And look at this,' he pointed to another figure, 'this is a man, with extended phallus, and here . . .'

'Looks like a rabbit,' I said, 'and a frog.'

'This scene is symbolic,' Memo continued. 'We don't know of what, but it has great meaning.'

'And think how much effort it took to get here,' I mused, 'and to make a mural here in the darkness.' For a second I was transported back 3,000 years, to a man standing barefoot where I stood now, illuminated by the flickering light of a flaming torch, risking his life a thousand times over to get here and get out.

Further into the cave it got even wilder, and even weirder. We rounded a corner in a craggy tube and came face to face with a tapir. A stone tapir. In 3D. The Maya had found a peculiar natural formation in the shape of an animal's head, and they'd scrawled on it in charcoal to give the impression of a tapir. It was remarkably accurate, with ears, nose and eyes in perfect position. And above it, emblazoned on the ceiling, was a parrot so perfect it looked as if it had just been stencilled by Banksy. And it was put there when my people were just stumbling out of the Neolithic.

After ten hours underground, we came into the final cavern of this arm of the cave. We all gasped as we walked inside, unsure if we were looking at an optical illusion, a natural formation, or some ornate work of art by the ancient Maya. There was a huge central pillar, and beyond it the ceiling was curious flowstone drooping down in rivulets. Or when lit right with the red and orange glow from burning torches, as the Maya would have seen it, it was a tree.

'I cannot prove it,' Memo said, 'but this to me is why everything else is in this cave. Why they put so much effort into getting here, why it was so sacred. This is the tree of life. It could have been the most important place for the Maya from this area.'

'The tree of life,' I repeated.

'Exactly.'

We all looked at the vast undercanopy in awe. It was simply spellbinding.

We were coming towards the end of the expedition, and we hadn't explored as much new cave as I would have liked. We'd talked over and again about how only a fraction of the caves of the Yucatan had been explored, but in reality we'd only charted and measured around half a kilometre of new passage. We needed a grand finale, a final push into the unknown darkness.

While I'd been in the interior with Guillermo, Robbie had been back in No Name cave, trying to find the way on. Unfortunately, after four or five frustrating days, he had hit a brick wall.

Literally. No Name was not going to give us our crowning glory. We needed something else.

Knowing what we were looking for, Robbie had gone back through his notes, and remembered a cave he'd found over ten years previously. It was a substantial dry cave system, with limited water deep inside. He had been intrigued, but couldn't spare the time and effort to get scuba tanks there. It seemed perfect. With just days left of the expedition, everyone packed up their gear once more, preparing to spend a few days out in the forest.

After yet another hot hike, we found ourselves standing outside the entrance to easily the most dramatic cave we'd seen yet in the Yucatan, forgotten in dense forest but no more than four or five kilometres from the road. It really was spectacular, with 270 degrees of cavernous entrances opening into grand chambers with stalactites as thick as an English oak. Though few people had been into the cave in recent times, it was obvious the place had a deep human history. The sandy floor was emblazoned with countless tracks from pig, paca and coati, all animals the ancient Maya had clearly hunted here. At the edges of each chamber, runways no more than a foot across had been constructed with paving slab-sized limestone blocks. As these were the only easy way for animals to move in and out of the caverns, the track down the centre of the runway was worn smooth by paw, claw and belly. My hypothesis is that the ancient Maya would have put basket traps at the ends of these runways, then frightened the animals out of the caves. They'd have sprinted down the runway and into the basket.

The coolest tracks, though, were located running in and out of sandy mammal burrows just inside the cave. A single thin track, like that made by a bicycle tyre, ran down into each burrow, and there was a twisting, turning track leading out. These were from a single actively hunting snake that had done the rounds of all the burrows looking for a meal, and it was big. Perhaps a boa constrictor, or a really big indigo snake (though they mostly eat other snakes).

Other signs of the Maya were even more dramatic. Close to one of the entrances was a stone altar, which had been packed out with crude stucco plaster, in which several carvings were still visible. One was a smiley face that could have been transplanted from a 1980s acid house T-shirt. Beneath it was a profile of a penis, complete with testicles, which could easily have been scrawled on a modern toilet door. There were shards of pottery everywhere, and countless stalactites that may or may not have been carved or shaped – after a while it was difficult to tell.

As we stepped inside, I found myself (for the thousandth time) starting to hum 'Going Underground' by The Jam. Within a few minutes, I'd ear-wormed everyone on the crew, and at regular points over the rest of the day someone would start whistling or humming it. (Could have been worse. It could have been 'Deep' by East 17. There's no way I'd want the soundtrack to my doom to be East 17.)

Just metres away from the light, the cave got really special. Rarely have I seen anything so decorated. The ceiling was hung with millions of stalactites, and throughout most of the galleries they were uniform, pointed like daggers dropping down towards us. It seemed to me the cave was an example of relatively new (geologically speaking) and rapid formation. In some places the calcite had formed around tree roots and encased them like a child's plaster of Paris school project. These formations could only be decades old, as opposed to the thousands-of-years-old speolothems I was more accustomed to seeing. Much of the rock had the consistency of bubbly fresh cement – crumbly and aerated. Whip spiders, crickets with enormous antennae and shield-backed cockroaches scampered about ghostly bat wings.

In one particular room, the ceiling was hung with helictites or 'eccentrics' – bizarre formations which had begun in a conventional way, then gone plain bonkers. Suddenly the calcite or aragonite starts to corkscrew off like curly fries and slinky toys. They're impossibly delicate, so are not found in caves where people roam, and don't last long even when the most careful and respectful of explorers come across them. I'd never seen helictites on this scale before. It was as if the ceiling were adorned with chandeliers designed by Doctor Seuss. Robbie and I gazed in utter delight at a glory of geology few human eyes had ever seen.

Eventually, though, we came down to the water. It was a good 300 or 400 metres into the darkness. The pool itself was no more than large hot tub-sized, with formations emerging from and dropping into the water, giving it the air of a fantasy fairy pool. A disgruntled-looking catfish with numerous barbels protruding from its pouty lips wandered past, as well as a few pedicure fish and some blind cave shrimp. It looked like some quicksilver phantasm might silently emerge at any second.

We geared up in the pool, moving carefully and cautiously to avoid kicking up silt and a replay of our debacle in No Name cave. What we were likely to find was to be as committing as cave diving gets, with a squeeze right at the beginning.

Beneath the surface, the water was shallow and the bottom composed of slivers of calcite like flat chalky potato crisps. Robbie ran out the line, and I followed, with Katy on my shoulder. Within minutes of leaving the crew behind, we swam into an optical illusion as bonkers as any halocline. The bottom appeared to be rolling sand dunes of calcite chips, the ceiling a perfect mirror image. As Robbie swam across the dunes, all of a sudden he also swam across the ceiling! It was a perfect reflection caused by the undisturbed water surface above. At that second, my bubbles broke the surface tension, and the illusion vanished.

This new room appeared to me to be blocked in, but in a far corner Robbie found a dune that didn't quite connect to the rock wall, and in fact hid a narrow gap falling downwards and onwards into oblivion. He signalled to Katy and me to wait, and dropped over the rim and was gone. Katy and I popped up to the surface in the tiny enclosed air bell.

'Where the hell has he gone?' I asked.

'With Robbie you never know,' said Katy, before taking a big breath. 'Be careful in here – I'm not sure how good this air is.'

'You're right,' I gulped. My breathing rate had started to increase, a sure sign that the air pocket did not have enough oxygen, and at best had a too-high percentage of carbon dioxide . . . potentially it could be even worse. The pulse quickened, the breathing increased even more. I thrust my regulator back into my mouth.

And that was when Robbie popped back up. 'It goes!' he said simply. 'It's a bit of squeeze down here below us, then it opens out. Are you up for this?'

'Hell, yes,' I said, with a confidence I did not feel. 'Let's do this.' I had never sounded more like a failed audition for an American action movie.

From the air bell, the floor dropped away steeply in a narrow slice, only just wider than my body. If it had been solid rock on both sides, I would not have been able to get through it, but the floor was calcite chips, so gave me a bit of wiggle room. I've never liked squeezes. On a cave exploration in Borneo I got properly wedged in a constriction, so much so that even a bucket of lard wouldn't have got me out. I was informed that I'd have to stay there till I lost some weight. Obviously in a flooded cave this wasn't an option.

Robbie had glided through like a greased weasel, but it was more of a battle for me. I found myself having to breathe out to push through. Once beyond the restriction, though, the slice became a tube, an arm span across and top to bottom, but bordered by fragile formations similar to those found in the dry cave outside.

It's difficult to say how far we swam down the tube, as we were focusing on steady movement and moderating breathing – and I was doing my best not to burn off my gas quicker than the rest of the team. My guess was we'd swum the length of a football field before the tunnel met a crossroads. We pitched straight on at first, but hit a dead end. We then backtracked and pushed left. Again it came up with a blank. To the right there was a gallery of stalactites, which looked like the bars on a Western gaol. Surely no way through there? Robbie gave us the signal to hold, then went for an experimental wriggle. Katy and I hung in the small tube and watched as Robbie rolled to one side, jiggled his tanks around and then started to squeeze his way through. A few more grunts and he was through and gone. We allowed ourselves to drop down onto the bottom and waited.

After ten minutes had passed, I checked my gauges and computer. We were ten metres underwater, and I still had plenty of air. I switched off my main light to save battery, relying on the residual light from Katy's camera, and closed my eyes.

This was taking a very long time. I checked my gauges for about the thirtieth time, adjusted every strap on my equipment and did everything I could to take my mind away from the reality of being entombed in a flooded tube surrounded by an infinity of rock. I was also starting to get cold. Having made such a mess of my temperature on the last few dives, I'd returned to a thinner wetsuit, figuring that slogging through dry caves would get me way too hot. Now, lying still on the bottom, even the warm cave water was starting to get to me. I was shivering like a dumping dog, wrapping my arms around myself and clenching all my muscles to try to stay warm.

'*Are you OK?*' Katy signalled me for about the twentieth time, clearly concerned. Even through my regulator she could see my shivering, and see I was in trouble. I wobbled one hand laterally: '*I'm borderline.*'

She spun one finger around on its axis. '*Should we turn around and head back?*'

I spread my hands in an unsure questioning motion, then tapped my watch: '*Where the bloody hell is he?*'

We waited just a few minutes more, but then we both hit our thirds. There was nothing we could do, we had to turn back. Knowing that most of our air had been sucked down waiting in this tube, we could probably have pushed it a bit longer, but how would that help? If Robbie really was trapped somewhere beyond the stalactites, there was nothing either of us could do to save him. There was no option. The decision made, I powered out of the cave full pelt, trying to get the blood back into my body and to stave the chill from my bones. We broke the surface together.

'What happened, Katy? Should we go back for him?'

'No, no,' she said, 'he'll be fine, he's probably just surveying or something – this is Robbie we're talking about here.'

I wasn't so sure, but my teeth were chattering like I'd been ski-ing in my Speedos. And we had no more air, so we couldn't go back. At that moment, light emblazoned across the passage behind us and shadow raced behind the speleothems. He was back! Robbie emerged from the water slowly and dramatically, his mask breaking the surface like something from a monster movie.

'What happened, Robbie?' I needed to know. 'Where the hell did you go?'

In answer, Robbie just held up his reel. It was empty. He'd let out 500 metres of line.

'It just keeps on going,' he said, 'and there are linear swirls of sand that shows there's current. There is a big cave off there somewhere.'

That night, we sat around the campfire and watched some of the footage back on Robbie's camera. Beyond the stalactites, the tunnels had got much smaller before they got bigger. In order to progress, Robbie had needed to remove both of his tanks and push them through the holes ahead of him, donking against each other audibly as he shuffled them along the floor. He had then proceeded to squeeze himself through a gap smaller than the inside of a Mini Cooper's car tyre, not much bigger than his own body. And there wasn't just one of these, there were five, one after the other. This was a level of commitment that I had never looked for on this expedition. I knew that 'tanks off restrictions' were a reality in cave exploration, but I didn't expect it would ever be something I would attempt myself. We now had some very hard questions to answer.

'So, have you ever taken your tanks off completely inside a cave?' asked Robbie.

'Of course,' I replied. This was sort of true. I had done so on a training course in a Welsh mine. 'I'm keen to give it a go. What's the worst that could happen?' I added.

'Well,' said Scott, never one to sugar-coat a turd, 'you could get stuck in a gap, half a kilometre underground, and we'll only be able to get your body back when the isopods have eaten enough of you that we can drag out the bones. We'll send Helen back your dive kit.'

'And I get to keep your houseboat,' added Aldo.

'I think we could get you through the first restriction,' said Robbie, 'and then take you to the really small one. You have a look, see what you think. And then we can have a look around, see if the cave goes off in any other directions.'

We were eating a dinner of boil-in-the-pot instant noodles with tinned vegetables.

'Who put Aldo in charge of shopping?' I asked.

'What's wrong with this?' he asked, 'I like Pot Noodle.'

'Exactly, for a Scot this is luxury cuisine.'

'Good thing about having a beard,' he replied, ignoring my attempt at needling him, 'is that I'll find some bits of spicy veg in there later on tonight, so I get to eat the meal twice.'

'If I die diving tomorrow, and my last supper has been a Pot Noodle . . .'

'Nobody's going to die tomorrow,' Robbie reasoned. 'We are well prepared, we go cautious, and if anything goes wrong, no matter how small, we turn the dive and come out. No problem.'

'But you are on your own,' Aldo said. 'I mean, I'm supposed to be in charge of safety, but as soon as you duck under that ceiling underwater, there's nothing I can do to help you. If there's a problem, your only option is self-evacuation.'

'Self-evacuation?' I blurted. 'How's that going to help?'

'It could make you lighter,' Robbie quipped, 'help you get through the little holes.'

We chatted on for several hours, fuelled by the two cartons of sweet red wine our team had been kind enough to put in the kit bag. After a while, Robbie got onto the subject of teaching cave diving and why it really wasn't for him.

'I mean, all the diving agencies, they all start from ocean diving and adapt for caves. They tell me I have to teach the buddy system, teaching people to share air with their buddy if there is a problem. I say this is bullshit, I refuse to teach them.'

'But surely buddy breathing's just one of the techniques you have to learn,' I said. 'I mean, not everyone dives completely on their own like you do. It makes sense to learn how to work as a team, doesn't it?'

'Whether you dive on your own, or whether there are ten of you in there, when you cave dive you have to look after yourself, end of story.' Robbie was animated now. 'I've had Navy SEAL divers come in and dive in cave exploration and totally freak out, and PADI instructors that I won't even let start the course, 'cos they think they know everything, and they don't know anything. Just this week I had some shit-hot diver come in, and five minutes into an easy cavern, they went into shock – throwing up, shaking, traumatized. I would rather take someone who has never dived before and teach them from the start only about cave diving. That is the way to be safe.'

The conversation continued in this vein for about 15 minutes. It was not an ideal exchange right before the most challenging dive of my life. I lay in my hammock that night, plagued with the potential and the problems. I had a restless night's sleep, tossing and turning, seeing myself in the house of daggers, with the ceiling descending on my head like some kind of trap in a B movie. I had a wife at home and a baby on the way. We'd already had a successful expedition, I didn't really need to risk everything. But on the other hand, beyond the restriction could be something remarkable, a cave passage of real significance that could take us to the next level. It seemed the only option was to give it a go and see what happened.

It wasn't until the next morning, when we had a more reflective chat without the wine, that Robbie's words really hit home.

'Cave diving is potentially the most dangerous of all sports,' Robbie said. 'If it goes wrong, even for a few seconds, then you die.'

'So what are the potential problems?' I probed. 'Squeezes? Running out of air? Losing light? Getting lost?'

'Your only enemy when you are cave diving is yourself,' Robbie replied. 'Panic is the number one. Every problem you get into, you can get yourself out of it, if you are mentally strong. But you have to get your head right. If you panic – even for a few seconds – then you die.'

'There's no injuries in cave diving, right?' Aldo agreed. 'It's binary – live or die. You either get out alive under your own steam, or you don't and it's body recovery.'

'Right,' said Robbie, 'but everyone I know who has died, it's been because they got too confident, or they lost control. You have to be responsible for yourself, you have to be focused on how to get yourself out of any problems you might have, and be 100 per cent self-reliant, or any tiny issue is going to kill you.'

This was why Robbie refused to teach buddy breathing. In all sports, experts talk about how important state of mind is. How you can improve your performance through controlling your mental state. I wrote in my book, *Mountain,* about how vital controlling emotions and fear are to climbing, and it's true . . . but not in the same way as it is for cave diving. I could be the most mentally tough human being on earth, but I will never climb like Alex Honnold, or whitewater kayak like Rush Sturges. However, in cave diving, your mind is the ONLY thing that is important. There is no reason why a four-foot tall, 70-year-old grandmother couldn't be the greatest cave diving explorer on earth if she had the mental fortitude.

As Daniel Kahneman explains in *Thinking, Fast and Slow*, our brains function in different ways under stress. Reactions in life-or-death moments are based not only on pure animal instinct, but deep training and experience. An experienced firefighter may recognize the signs of a flash fire a second earlier than a civilian. A Special Forces soldier might duck before a sniper's bullet is even fired. These reactions are the difference between life and death. In cave diving, the worst-case scenario is for your air to stop. Nothing creates panic faster than not being able to breathe. You have two or three seconds at best before the red mist of panic descends. If you're cave trained, that's enough time to switch to your backup system, or to reach for your tank valve and check it's not been knocked off. However, if your animal brain is trained to instantly look for the help of your buddy or to swim for the surface, you are finished.

As the sun cut light beams down through the smoke of our campfire, we boiled up our last cup of crappy instant coffee and feasted on a breakfast of granola with powdered milk. This was one thing I knew I wouldn't miss from this trip. Breakfast is my biggest meal of the day – I like to dine on a banquet of roasted vegetables, green smoothies and the hipster covenant of poached eggs on avocado with Barista-quality coffee beans making the perfect flat white. For the next year of expeditions, however, I knew it would be Nescafé and porridge, plus a punishing amount of cheese. There was, though, enough caffeine to get me out of my hammock and into my wetsuit.

The mood at the waterside was sombre. The team had all been talking in hushed tones about the footage on Robbie's camera, how what he did was 'insane' and made them 'feel so sick I could barely sleep'. For those in charge, like Scott and James, there was probably a real feeling of responsibility, of how culpable they'd be if something went wrong and they hadn't stopped me. For the others there was just the nightmare of contemplating how they'd feel if it were them going into those squeezes.

So it was a quiet and focused team that prepared for the day's diving. Katy placed two lights into the edge of the pool, and it glowed into life, shimmering with an eerie ethereal light. The light should have made the water more appealing. Instead it looked like a portal to another dimension. And we all know how well that usually ends.

It's an odd experience standing in your pants deep, deep underground, holding onto one 10,000-year-old stalagmite to keep your balance, draping your stinking wetsuit over another, using your head torch to scare off the shadows while you spit in your mask, after first flicking away the cockroaches and cave crickets taking up residence in your fins. 'Process' in cave diving is your friend. While you're thinking about where to seat the weights on your belt, or how to make sure your mask doesn't fog in 100 per cent humidity, you're not contemplating the essentially foolish escapade of deliberately swimming into one of the world's most hostile environments; a place where there is none of the most important element for human life, oxygen. A place where without a functioning torch you'll definitely die. The less you think about all that nonsense the better.

We were still several months before the news story of the stranded boys' football team in Thailand brought cave diving thundering onto the global scene. Precious few people in the whole world would have any understanding of what I was about to do, and appreciation of why. I was beginning to wonder myself. But it was time. Bubbling beneath the surface, I waited for the sediment to settle. The line stretched off into the blue. Robbie's fins faded. We pushed through the original restrictions and chambers, moving as if in a trance to keep heart rate down, breathing dampened, saving air for the fear ahead.

We found our way back to the House of Chill, where Katy and I had lain shivering, waiting for Robbie's return, and then into an increasingly tangled web of cave formations. How and why would anyone even think to look for a way on through here? We'd swum past a dozen such places and not even stopped to investigate, but somehow Robbie, with his decades of experience, knew this was the way on. Perhaps he sensed the faintest of flows agitating the hairs on the backs of his hands, or the faint movement of silt, or the sway of the line . . .

The line led into a maw of stalactites, fragile strawers and bamboo-like stems, floor-to-ceiling. It was like doing an army assault course, wriggling under barbed wire and through tyre tunnels, but while wearing a spacesuit. The gap through the stals was small, not much larger than my body, but it seemed I could slither through without having to take off my tanks. Twisting my body left and right, I made slow progress, centimetre by centimetre, dragging myself along the bottom, bumping my head on the roof right above me. The dong and bong of my tanks hitting rock echoed through the water, a minimalist orchestra of cowbells.

And then I stopped. I couldn't pull myself through, something had snagged. I tried to reach back down my side to find the problem, but the gap was too tight. When I edged myself backwards, inchworm style, it caught again. Now I could go neither forwards nor backwards. Robbie and Katy were both through already, and the water was starting to turn murky. I couldn't move. My worst nightmares came rushing in on me. Katy's camera lights momentarily blinded me. I could see the condensation trickling down the inside of my mask, and hear the blood thumping in my ears. Then Katy swung her lights away, and I could see her face, connect to her human eyes through the gloom. She held up a single hand in the manner of a policeman directing traffic. *Breathe.* Putting her camera down, she slid one hand down the side of my tank. I leant over to the other side, and my weights plopped me down with a heavy thunk.

This is what a goldfish looks like just before it gets plucked from its tank and flushed down the toilet, I thought to myself.

Katy shook my tanks, rummaging around for whatever it was that had got caught. Yes! She'd got it! Sliding forward, I popped – champagne cork style – out into the next section of passageway. My nemesis had been no more than a cable tie that had looped itself around a nubbin of rock, but they use those things as handcuffs – it would never have broken and I couldn't have got to it myself. It brought into sharp focus the commitment Robbie makes every single time he dives solo.

Beyond this tricky squeeze, the corridor continued unabated in a lengthy restriction. While it may not have had the grandeur of the bigger caverns, it was undeniably beautiful. The gap marked a bedding plane, a joint between two different layers of rock. Bedding planes in limestone are inevitably the weakest portion of the substrate and when water makes its insidious passage, it tends to erode along these lines of weakness. Robbie, with his explorer's eye, had discerned the difference between the strata and therefore the most likely way on into the bowels of the earth.

We twisted on through several more restrictions until we got to the final point Robbie had reached the day before. I had been anticipating this moment with utter terror. The point where – half a kilometre from air – I would have to make the call whether to remove my tanks in order to proceed. In my waking hammock nightmares, I'd run through what I'd do here over and over. The stern talking-to I'd need to give myself. The horror it would cause me, the exhilaration I'd feel when I conquered my demons. It had all been about this moment. And now I was here. And none of that mattered.

The gap ahead of me was not for mere mortals like me. Forget diving, tanks and the fact that we were underwater at all. I could not have got through that gap even if I'd been stark naked and greased in goose fat. (I apologize profusely for putting that image in your head.) I couldn't have got through that gap at any time since my fifteenth birthday. I made the tentative 'maybe' signal to Robbie, but really there was no need to contemplate. It was a definite 'no'. Robbie could go on, Katy maybe, but for me this was as good as a solid dead end in the cave: I'd reached my limit.

Robbie and I smiled at each other. For the first time I managed to actually shake his hand rather than his fist. He must have known that this was beyond me, but had brought me here anyway to make my own call. I had to be responsible for my own fate.

Many hours later, back in sunlight with the cave's darkness already an ancient memory, we sat swatting the incessant mosquitoes and pondering the surreal experiences of the last six weeks. My limited ability and experience had held us back, but not prevented a voyage of discovery into a wonderworld more beautiful than I could ever imagine.

When the story of the boys trapped in the caves in Thailand broke some months later, and cave divers tried to figure out how to save them, I felt my experiences in Mexico had given me a particularly stark perspective into their enterprise and their heroism. They swam through caves like ours, yet in zero visibility, with the whole world watching, and bearing the lives of their young charges with them. It is truly humbling.

But perhaps even more humbling is the sense of what lies hidden here beneath the Yucatan. The miracles and Mayan marvels yet to be discovered. In some ways this project had been *Expedition* in microcosm. It provided us with conclusive proof that there were treasure troves waiting to be found, and that with the right teams, we could be the prospectors who reaped the gold rush. We'd made a start on a very big year.

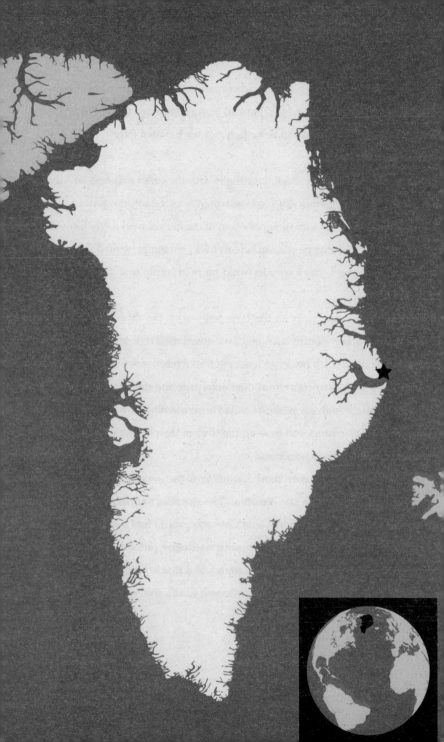

3. FROZEN FRONTIER

ARCTIC GREENLAND

He raises his snout onto the frozen wind. At this distance I can see the gloss on his black nose, the steam as his warm breath hits ice air. I can hear his breathing, see every shaggy hair hanging down over his curved claws. It honestly feels to me as if the ground shakes with his every half-tonne step. That, though, is a trick of awe. Even the biggest, heaviest polar bear can walk on thin ice that I'd crash straight through. His step is soundless as blancmange dropped on a heavy shagpile carpet. At a distance he looks unblemished – pure white perfection. Up close you can see the ugly diagonal scar across his snout, the mark of a confrontation with another male bear. All of this would be the greatest thrill in the Arctic . . . were it viewed from the safety of the deck of a mighty icebreaker, or with the vibration of a snowmobile throbbing on my bottom, safety just a rev and a roar away. But here, out on this icy beach, with nowhere to run, we are in his house, in the lair of the bear, the domain of the biggest predator to walk the earth. He is no more than three strides away, and I can neither run nor hide.

The frozen regions of our planet are called the cryosphere and they are the most treasured wilderness to any adventure seeker. Nowhere has so much impact on my mood, such an abrasive, cleansing action on my soul, as the high Arctic. The quality of the air is ethereal. It is so clear that it's sometimes difficult to tell if an iceberg or glacier snout is a couple of hundred metres distant or 20 miles away. The cobalt blue of the skies, and the semaphore signals of the clouds, herald the incoming weather before the weather satellites ever can. And the wildlife . . .

In the summer, it can be alive with birds, fish and whales in a way that few other environments can match . . . or it can be so utterly devoid of life that a single midge or lost gull becomes an object of fascination. Sometimes the wind cuts through you with such ferocity that it seems to be trying to chill your intestines. On the same day, the solar glare will flay the skin from your face with the intensity of an industrial furnace. It is the wildest, most untameable place on earth, and I adore it.

But although we humans have tried and failed to tame the Arctic, we have succeeded in screwing it up in ways too immense to contemplate. The greatest of those ways is climate change.

An ice-free Arctic Ocean will happen sometime this century, and a summer where the North Pole is ice-free is predicted by some climate scientists to be possible any year now. The Arctic is not a continent like its southern cousin. It's an ocean covered with ice, and in fact it doesn't get quite as cold as Antarctica. The thermal properties of the water mean it stays a little warmer, though don't get your bikini out just yet – it can still get down to minus 70 degrees centigrade, and *average* temperatures are well below freezing.

I made my first television series on climate change back in 2006. We talked about the Arctic as being the 'canary in the coal mine'. The glaciologists, climate scientists and biologists I had the great privilege of working with made ominous projections, and talked of feedback loops and cascading temperature increases, predicting that the next decade would be the hottest on record, with each year getting successively hotter; that sea levels would rise by over a centimetre; that dramatic freak weather events would increase in frequency. Every single thing they predicted happened, and then some. And yet still there are people around the world who peddle the myth that climate change is not real.

The Arctic is still the canary in the coal mine, the place where climate change is something you can see and touch and feel. So I wanted to come up with an expedition that had old-fashioned expedition objectives, but along the way would investigate the reality of how the Arctic is changing.

It's tricky to come up with true firsts in the Arctic. It's an open environment, without caves, canyons and hidden gullies to explore. Explorers have been coveting its grand firsts for generations, journeying across the poles in ever more ambitious and stripped-back suffer-fests. Yet they are just following in the footsteps of Arctic peoples, who have been traversing its waterways and land masses for millennia. To my mind, the biggest first that remains to be done is to sail or kayak across the Arctic Ocean, on the first year when it is completely free of ice. This, sadly, is a first that will become possible in my lifetime. It could be as soon as 2020. However, it won't be possible this year, and is a mission with a timescale and budget beyond my scope . . . for now.

There are, however, other adventures that could well be possible this year for the first time in human history. One of those was to attempt to sea kayak down the world's longest fjord in spring, in June, when until a year or two ago, it would have been totally frozen over, and impassable to any water craft of any kind. On 17 March 2018, the Arctic sea ice was at its fullest for the year, and yet this was the second lowest in recorded history. The lowest extent was this same time in 2017.

This was the most critically timed of all the expeditions. We had to be here as the early pack ice began to break up, but this meant being out on polar seas perilously close to Helen's due date. The troubles we'd been through meant she could potentially be rushed into hospital at any time. And once I was out on the Arctic wastes, it would take me at least a week to get home.

Before I left, we went down to stay with Helen's family in Cornwall. A heavily pregnant Helen and I walked down to Logan Rocks, a dramatic promontory looking down onto Cornwall's most beautiful beach. Huge waves smashed against the granite, and choughs and fulmars soared about us as we sat silently, hand in hand. There was no doubt that this would be the hardest trip for both of us. Saying that Logan Rocks would always be our happy place, we promised each other that anytime we felt a bit lost or lonely we'd come here. Helen literally, me perhaps just in my head.

Greenland is the world's largest island, bigger than the UK, Spain, Italy, France and Germany combined. However, in that whole vast land mass, there are only around 56,000 people. Fewer than live in Maidenhead. That makes it the world's least populated country. Eighty per cent of that land mass is the ice sheet, the biggest outside of Antarctica, and is itself seven and a half times the size of the UK. This 2.4-million-year-old ice sheet is melting faster than any other part of the planet, twice as fast as anywhere in Antarctica, and will go disturbingly soon. The melt of the Greenland ice sheet contributes 1.2mm a year of sea level rise – worldwide! When it does go, global sea levels will rise by six metres, and one billion people who live in the flood zone will have no homes.

Almost all of the few folk who live in Greenland are on the west coast; the east is just empty. The largest national park in the world is here – a wonderland of mountains, tundra and frozen seas that is home to the pure white gyrfalcon, the Arctic wolf and the Beluga whale. We were heading to Scoresby Sund, which is a gigantic fjord that penetrates 110km into the eastern coast. The feeder fjords may go twice as far inland, making this officially the biggest fjord system on the planet.

The team for this one was just as specialized as we had assembled for Mexico. Keith Partridge has done most of my gnarliest expedition films, and had been on hand to capture me being flushed down an Alaskan glacier, tumbling off a Venezuelan tepui and nearly getting swallowed by a pod of 40-tonne humpback whales.

I'd just worked on a film with John Livesey on the north face of the Eiger. He's a Scotsman with the work ethic of a leafcutter ant and the brain of a physicist, who has achieved game-changing things with mini-cams, which could potentially make a huge contribution to our ability to film the expedition.

And then there was Sarah MacNair Landry, who would be our camerawoman when the kayak team separated from the main crew. When we started putting the team together, all I knew about her was the fact that on an Arctic expedition she had broken her back . . . and kept kayaking. She was immediately impressive, fresh-faced and blue-eyed, with a hugely energetic demeanour. Sarah lives in Arctic Canada in Baffin Island, runs her own dog sled team and adventure company, and, at 32, has more expeditions under her belt than any of us gnarled, bearded, horny-handed old-timers. She has completely dog-sledded around Baffin Island (in winter) and paddled first descents way up in the Arctic Circle. Not surprisingly then, she took everything in her stride and clearly had nothing left to prove. (Again, at 32.) She also has a wicked sense of fun. On day one, she deliberately mixed up the labels on our powdered foods. Keith put salt in his coffee instead of sugar. I went for powdered milk and whitened my own precious coffee with mashed potato. How we laughed.

Our gateway to the wild wastelands of East Greenland was

Ittoqqortoormiit, a remarkable town of 360 souls, clinging to sloping hillsides that drop down to a curved bay in Scoreseby Sund. The houses are clad in weatherboard and wooden planks, all painted bright blues, reds and greens, faded by solar glare and the recently receding northern winter. Snow was still piled high alongside the roads, locals gunned about on ATVs, and the streets steamed as puddles evaporated in the remorseless sun. The town smelt of hot wet cement.

This is the only settlement on this part of the coast. We could be convinced of our isolation, as there could be no one for thousands of miles who did not pass through here or nearby Constable Point. With no one else travelling out into the fjords or peaks as we would be the latter phase of the expedition, once there we would be assured of being literally the only people in a landscape bigger than the UK.

Usually on expedition, I do my best to learn as much of the local language as possible, even going to the lengths of getting near fluent in conversation. This can be one of the most dramatic leaps in breaking down boundaries with local people, and gives you access to culture and friendships that would otherwise prove elusive. To these ends, I quizzed the Inuit lady who was to be our local contact.

'What's the local word for "hello"?' I asked.

'We just say hi,' she replied. This was going to be a breeze.

'And how do you say the name of this village?'

'Ittoqqortoormiit,' she replied. It sounded like she was mimicking a toddler's table tennis tournament with her mouth.

OK, let's start with something more simple, I thought. 'How do you say thank you?' I asked.

'Nan tuk at,' she replied.

I tried to mimic her. She sucked in her breath with shock. 'Don't ever say that!' she gasped. 'That is a very dirty word.'

Suffice to say, I don't think I will be getting fluent in Inuit on this trip.

Wandering down to the guesthouse I almost instantly came face to face with a musk oxen skin stretched out and drying in the sun, the straggly long hairs of the pelt nearly half a metre long. Near it, on a baby-blue wooden fishing shack hung with rustic fishing nets and paraphernalia, were two ivory-white polar bear pelts, stretched out on racks. One of them was obviously from quite a small bear, potentially even a young animal.

The polar bear is known by Inuit people as nanuq, which literally translates as 'worthy of respect', the ever-wandering one. They are an icon of the north, with a lonely life of starvation interspersed with occasional gluttony, in an environment that could not be more alien to our own. *Ursus maritimus*, the maritime bear, is the undisputed heavyweight champion of this white world. While there have been individual brown (grizzly) bears from Kodiak Island in Alaska that have equalled their size, the polar bear is the largest land carnivore in the world. Males can weigh in at three-quarters of a tonne and stand two and a half metres tall. They have been seen walking in a dead straight line for 20 miles towards food they can only have smelt. From 20 miles away. Despite their energy-efficient loping gait, they can record extraordinary explosive speeds over short distances, and are absolute experts at not being seen. They are also one of the few mammalian predators who will deliberately hunt, kill and eat a human.

Bears are a game-changer in this part of the world, and define almost all of your actions. While more than 99 per cent of human interactions with polar bears will conclude with the bear beating a retreat, the exceptions are real and horrific. Usually these involve third-year male bears who have not long been driven away from their mother, stranded on shore as the sea ice retreats, starving, skinny, desperate. Or females protecting cubs. Or dominant males that start inquisitive and decide to take the opportunity of an easy meal.

With colossal home ranges, a bear could be literally anywhere, so you need to act at all times as if they are there. This meant we had to carry a 12-gauge shotgun filled with slugs that would stop a charging rhino, personal alarms, flares, and a bear 'fence' to put around camp. Despite all that, if we were ever making camp, someone would have to be awake and on bear watch, rifle always ready. This wouldn't be so much of an issue if we were close to the film crew, but when we were remote from them, it would mean the four paddlers would have to be doing two-hour shifts through the night, sitting awake and not moving in the worst Arctic chill, even through the graveyard shift of 3am to 5am, when the human body is at its lowest ebb.

Sarah, as an Arctic veteran, was instantly grumbly about the whole bear-watch thing – despite having had scores of close calls with bears in the past.

'I reckon that Aldo should do the watches,' she said. 'He's had all that experience as a sniper.'

'Do me a massive favour,' I asked. 'When he gets here, will you only talk about his "army" experience? He'd love that.'

'I'll do that,' she said, 'if you'll take one of my bear watches.'

From there on in, every single thing that any one of us wanted anyone else to do was offered at the price of a trade. One favour, for a bear watch.

The animal I most wanted to see, though, was the narwhal. It is one of three Arctic species of whale, along with the giant bowhead and the white whale or beluga (these are the three that *only* occur in the Arctic – there are lots of other whale species that can occur here, but are also found elsewhere). I've never seen a narwhal, despite many months at sea in the Arctic, and it has become a holy grail for me, up there with the snow leopard. Now we were here, in narwhal territory, and at the perfect time of year, as the ice started to break up. Great! And there had been sightings already.

'We saw eight narwhal just last week,' one of the locals told me at the helipad.

'Really?!' I responded, ecstatic. 'Where were they?'

'Just out in the Sound,' she responded. 'We managed to shoot six of them, but the others got away.'

★

'I just saw a swallow,' said Rosie, our director, as we were waiting for the heli to bring in the first load. It was one of the first signs that things might not be as they should.

'Nah,' I said. 'Maybe a shearwater.'

'It had a forked tail,' she responded.

'Maybe an Arctic tern?' I offered.

'They're bloody white, Stevo!' Now she was really pissed off. 'It was definitely a swallow.'

'What are those small sea birds called then?' Sarah asked.

'Storm petrels,' I added helpfully.

'Yes, that's right,' she said. 'Probably storm petrels, Rosie.'

Rosie simmered. Ten minutes later, two swallows flew past.

'Those are swallows,' I pointed out. Rosie looked as if she might force-feed me her binoculars. Rectally.

It was a conundrum. What on earth were swallows doing here at all, let alone at the beginning of June?

The swallows were not the only bird we saw standing up on the promontory above the town. A pair of snow buntings cavorted before us, the male in his plush breeding livery of purest white breast and head and glossy black wings. These are the most Arctic of all songbirds, the most northerly breeding passerine or perching bird, and the only really omnipresent small bird. The male flitted about, finding the most prominent rock to perch on and let rip with his simple song. All seemed a little bit pointless, though, as there was only the one female about. Competition was not exactly rampant. I mean, he could probably have got lucky without all the effort.

Snow buntings were to become our nemesis. They were metres away from us all the time we were in Inuit settlements . . . until we set up the camera on a tripod.

'I bet you one bear watch you can't get a shot,' I challenged Stuart, our second cameraman.

'You're on.'

For the next few days, the snow bunting would sit on the highest, most obvious rock possible, until the second Stuart had his huge camera lens, tripod and camera in position. Then it would fly off. Despite their glossy plumage and sweet song, they have macabre reputations. With nesting sites few and far between, they'll make their home just about anywhere. They've more than once been seen nesting inside a frozen human corpse.

We needed to kill a couple of days before Aldo Kane turned up, so we decided to go and take a good look at the pack ice. The early melt meant it would be hard going, but there were a few sledders who were still running dogs, so we took three teams and headed up to the top of the ice cap to get a good view.

I don't need much of an excuse to run with dogs, as it's one of my favourite things in the world to do. I love the energy of working dogs sprinting through snow, tongues lolling to one side, white spray erupting from around their huge furry feet. However, I had never run with Greenland dogs. These are a far cry from the huskies I've run with in Alaska, Norway and the Alps, who look like they've just swaggered off the awards podium at Crufts, blue eyes glinting, glossy fur, almost too perfect. The Greenland version looks more like you've taken a couple of wolves, made them fight for a few months, then rolled them in tar. Of course, it's not actually tar, it's congealed seal fat and years of matted shit. My first move was to walk towards one friendly-looking dog, who was rather less crap-covered than his compatriots.

'Would it be OK if I said hello?' I asked their driver.

'You can try,' he nodded back to me.

'You're not going to bite me, are you?' I said confidently, going to ruffle his ears.

Without hesitation, he sunk his teeth into my arm.

Most of the dogs were wild and wary, snapping at each other's flanks, flashing teeth, meting out discipline to their subordinates. Their communication and interaction was pure wolf.

My Inuit driver Aqqalu was about six foot three, face stained creosote brown, crooked teeth flashing brilliant smiles.

'Are any of your dogs more friendly?' I asked.

'Yes,' Aqqalu said, 'lead dog is friendly.'

We walked down the line to the head dog. Dark in colour, he was furry as a bear. He was also the most terrifying-looking dog I've ever seen, muzzle criss-crossed with scars from battles with his team, one eye torn downwards and blood-red lips that curled to reveal brown broken teeth. He must have been coming to the end of his years as a running dog, but as a lead would be so valued as to be given every chance to run out his tenure at the front of the pack. I'm a dog person, I adore canines of every variety, and in general dogs like me too. This bruiser, however, had me nervous, and that's the worst way to be with any animal. As Aqqalu approached, though, this giant black hell-hound dropped down onto his belly, practically grovelling before his master, then leapt up to grab him and lick his hands.

'What's his name?' I asked.

'No name!' Aqqalu laughed. Seeing that I was trusted too, the lead dog repeated the friendly gesture towards me, leaping up and pretty much putting his feet on my shoulders.

'Can I pet him?' I asked, literally once bitten, twice shy.

'You can try,' he said.

Mentally preparing the evacuation procedure for when he mauled me like a lion on a baby gazelle, I ruffled his scruff and tattered ears, enjoying the contact, but certain I would never, ever turn my back on him. He could have snapped my neck like a twiglet.

The rest of the dogs busied themselves in scent marking everything that was placed on the ground anywhere near them. One of the cameras got a hot dousing of yellow slurry, someone's backpack was gloriously christened, and then in a moment of pure slapstick comedy, a dog snuffled up to Stu while he was engrossed in filming. We all watched with bated breath, none of us warning him while the dog lifted his leg and filled up Stu's boot. He didn't even notice until his welly was overflowing.

As I wandered down to check out Sarah's team, a bundle of fur came bounding through the snow towards me. This one was instantly giving out the best possible signs: ears up, tail wagging, gambolling puppy-like movements. The big husky leapt straight up onto me and rolled me into the snow, encouraging me to stroke her tummy, gnawing at my fist in the same way as my own puppy back home. This was a young dog, no more than a few months old, hanging with the main teams to learn her trade. I laughed like a drain . . . until I breathed in through my nose. The smell was like a long-drop toilet on a hot day.

For the first few hours we thundered up a rolling hillside, shrouded in thick fog that froze our ears and fingers. Then abruptly the fog burned away, and blue skies began to appear above us. Within minutes it became clear that there was a classic temperature inversion, where the temperature gradient has flipped, a cold layer of air trapped below a warm one. The cloud had formed low down in the valleys, leaving the sky above clear. Below us was a sea of cloud rippling out to the horizon, the highest mountain tops floating above it, as if we'd ascended to a celestial realm. Sarah was the only one who had the foresight to apply sunblock, the rest of us frazzled like rashers on a griddle.

When we came to steep hills, I had to leap out alongside Aqqalu to push the sled, taking the weight off the dogs as they thrashed around in the heavy snow. After a while, Keith the cameraman jumped off so we could get some drone shots. For the next half an hour the dogs dragged us forward, the sled running easily through the ice. Aqqalu turned to me and spoke the only words he ever said to me without prompting: 'Cameraman is very heavy.' Then he smiled and turned back to his team.

In Greenland and in much of the Arctic, the Inuit drive their dogs in 'fan formation' – every one on a separate line, the lead dog taking the apex of the arrowhead, often with a second helping them out. Strong 'wheel' dogs are behind them, forming the engine room of the team. The bit I found new is that there is also an alpha male or boss dog and alpha female, which is determined within the pack. This is not always the lead dog, which kind of blows my preconceptions of pack dynamics.

In Sarah's words: 'A good boss dog will put the dogs in their place, discipline some, break up fights. The driver can decide who the lead dog will be, the dogs decide who's the alpha. Buying an adult dog and introducing it to a team is nearly impossible. They're such a pack, they need to be socialized and brought up together. It can be done, but I would expect fights. They have to decide where they are in the pecking order.'

This formation is super-effective in the open terrain of the Arctic, but doesn't work anywhere there are trees, as the dogs are too easily tangled. The dogs seem to be constantly on the toilet, barely breaking stride to paint the snow yellow, or kick up black nastiness into the air behind them. At rest they're either sleeping, or fighting, snarling and howling. The second they take flight it is silence, every ounce of breath and energy being given to the chase.

These Greenlandic huskies were brought out here at least 4,000 years ago by people as working dogs. They have been bred for millennia to run through the snow, up to 100 miles in a day, towing loads, running for hour after hour. When you go with the driver to the groups of dogs, carrying harnesses, choosing those who will get to be in that day's team, the noise is phenomenal. Every dog is in full voice, leaping up, going crazy with excitement. Every fibre of their being screams 'Take me, take me!' Those who are not chosen howl with fury, ripping at their collars and battling to join their teammates.

Our puppy ran with us for the entire trip. It wasn't collared or tethered into the team at all, and was never encouraged in any way. Yet the young dog kept pace with us (and only our team), leapt in between the traces trying to take its place there, and yipped and yapped with enthusiasm. When she got out of line, an older dog would wield discipline with a snarl and a nip. At one point, puppy mistimed her run and went right under the runners of the sled. She yelped in pain and fright, and I screamed with horror. A British pet dog would have had its spine snapped in two, but our puppy leapt straight back to its feet, and just kept running, never so much as breaking stride.

Finally, after a good six hours thundering through the snow, we arrived at our destination of Kap Tobin (now known as Uunarteq). This small hunting settlement is right at the end of Scoresby Sund, where it is at its widest. With the same faded wooden buildings as Ikktoq, it's a choice hunting location, where spotters can see anything that travels in and out of the fjord. However, it was deserted and has no roads, so you could only really get about by snowmobile.

We were to stay at one of the cabins with the sled dog drivers, and in our heads had prepared to chat with them there about their changing lives. Once we got inside, though, it was not exactly what we'd expected. It was decorated like a 1980s tenement that had been uninhabited for a few years. Despite the total lack of atmosphere, it was a great little vantage point to watch the sea. We trained our binoculars out to the shifting ice and lumbering bergs, desperate for a sight of a bear or whale, and revelling in the abundant bird life. Scores of tiny little auks streamed across the fjord. They're the most numerous of all birds within the Arctic, with several million breeding pairs here in Greenland alone.

Far heftier were the flocks of eider ducks – the drakes startling in their black and white monotone livery. While we watched, they dived down to the bottom to gather molluscs, which they devoured when they returned to the surface. In between meals were bouts of calling with their curious 'ooooo', a sound rather like a disapproving character in a cheap farce. In amongst them we spied a king eider, which is a far greater prize. The male has an extravagant orange bill, and a generous shield tops his zinc-coloured head. Apparently, this shield is a delicacy for Arctic peoples, who will bite it off first upon catching a king eider.

Aqqalu took me out to the east of town, where the open ocean and fjord meet. For now the sea was still fast with pack ice, though there were patches in the clouds above that told a brighter story. 'Ice blink' is the name given to cloud that seems to be glaring, as if lit from within. This cloud sits over big areas of ice, and the degree of sun reflection from surfaces (called the albedo effect) means they blaze in the sky.

The contrary effect is called water sky, which is dark streaks across the clouds. It's kind of like 'the *Top Gear* filter', where the sky in a photo is black and ominous. This is caused by open areas of dark sea below the clouds, with a very low reflective index. This water sky has been vital to polar explorers throughout history, who will avoid it if on sleds or skis, but bear towards it if in kayaks.

Most first-time visitors to the Poles will be thrown by the other optical effects – parhelions and spectres, mirages and heat hazes. It seems counterintuitive that this, the coldest of all environments, should have blazing hazes that are more dramatic than in the Sahara. Mirages, though, are caused when temperature changes drastically with distance from the earth's surface. These different heat layers of air cause light to be distorted, while shimmering effects are caused by turbulence. Because you can almost always see to the horizon in the Arctic, and the heat gradients from the surface can be so dramatic, there is no end to the hallucinations the traveller can experience. Clouds become vast icebergs, icebergs become ships, blank horizons are suddenly monstrous mountains or raging dark hurricanoes.

Polar explorer Robert Peary famously discovered a non-existent island in the form of Crocker Land. This was almost certainly due to one of the most dramatic kinds of mirage, the Fata Morgana. Named after the Arthurian sorceress Morgan le Fay, they often manifest as land at the horizon, and were once thought to have been placed there by the witch to tempt sailors to their doom. Here in the Arctic they take the form of fairy castles or towering distant cliff faces, and feel tangible enough to bamboozle the most experienced polar hand.

That evening, we sat around in the hut/squat. It was cold and windy outside, and so we finished work around 9pm, with the sky still as bright as noon. For the first hour or so the crew busied themselves fiddling with camera kit, then people sat around and picked up their phones. We had no signal, but it has become as much second nature for us to pick up and start swiping as it is for the huskies to run with the sled. The apps and widgets soon outran their fascination, though, and the phones were discarded.

Finally the solitary silence was broken and we all started to talk. I didn't really notice the shift until I went out of the room to grab something, and came back in to find six people jabbering away as if the dyke had just burst. The stories came thick and fast. Sarah told of being inside when a polar bear 'ninja'd over the bear fence' and jumped on her tent, Rosie talked about socializing lowland gorillas in the wilds of Gabon, Keith about being trapped in a snowhole for a month with a climb partner who lost his mind. And if we'd had a phone signal everyone would probably have been on Candy Crush, or whatever the Polar version of 'Tinder' is.

The next morning, Aqqalu took us to a hot spring on the edge of Kap Tobin. It was scenic from a distance, with the steam coursing up in tendrils amongst the ice, but less so when we got close. Local hunters were using the main pool to clean off their prizes. Within the steamy puddle was a walrus skull and five polar bear skulls, the bubbling water boiling the skulls clean. The biggest bear skull was the largest I've ever seen, big and heavy as a bowling ball, with canine teeth stout and scimitar sharp. The browning bone and stench of rotting flesh made it even more intimidating. The grimness was offset by one long curved bone as long as my arm. I knew what it was, but wanted to know the local name.

'What do you call this?' I asked Aqqalu.

'It is penis,' he said. This much I knew. The baculum or penis bone of the walrus is notorious in anatomy. Most mammals have them, but the walrus has the biggest. So much so, they have long been used as clubs by Greenlandic bouncers. Imagine the shame of not being allowed into an Arctic nightclub and then being clouted by a seal willy.

On the way back, I jumped aboard with one of the other sled drivers. He hadn't spoken a word since we left town, and we'd assumed he didn't speak any English. However, after a few questions, it turned out that his English was the best of everyone's.

'This is not my dogs,' he confided. 'They belong to my uncle.'

'Really? But you run them really well.'

'Yes. I always have dogs,' he responded, 'but now is very hard.'

'Hard to learn?'

'No, hard because everything is change.' He gestured out to the open sea. 'Five year ago it was ice here until end of July. This year it was open from February. The time you can run dogs is shorter and shorter every year. Now I cannot afford money to feed dogs, so I have to give them away.'

This tallied with something Sarah had been telling me. Everyone assumed that the arrival of snowmobiles in the Arctic would kill the ancient art of dog sledding. But snowmobiles are expensive, fuel is expensive, and not everyone can afford them. However, if you have dogs and a rifle, you can hunt seals and you can feed the dogs. They reproduce themselves, they keep you safe from bears, they run for free food. But as the winter gets shorter and shorter, the season when they have to be fed but can't work grows.

'We can only run dogs here for one more week or two,' he said. 'When I was a child, we ran our dogs for months more. People who have dogs is less and less every year.'

It was an element to climate change that I had never even thought of, but one which is extremely poignant to me. The partnership between man and dog is a beautiful thing in this wild white world. It is a covenant forged over more than 4,000 years. The thought that it could come to an end in my lifetime, because of the profligacy of my kind, makes me want to weep.

I was shaken from this sad reverie by the driver asking me a question.

'Can I interview you now?' he asked.

'You what?'

'I have been watching you on the television. Can I film you, please?'

He then proceeded to take my camera and interview me on my opinions and views on his world, while the white scenery flew by, the dogs trotted in the thickening slush and the skies darkened above. It was a surreal and memorable moment in a very strange day.

The following day, Aldo turned up at Ittoqqortoormiit, He had spent the previous ten days living in total darkness in a nuclear bunker, deprived of natural light, timekeeping or stimulus, in an attempt to drive his circadian rhythms crazy for a television science experiment. No more than 24 hours after getting out of the bunker, he found himself in the land of the midnight sun. There was an even wilder look in his eye than normal. That night he bunked in with Keith and me, and at about 2am sat bolt upright.

'Where did I pack the big sharp knives?' he shouted, before falling straight back to sleep.

As if it wasn't sinister enough being on expedition with someone who is a highly honed elite forces fighter, we had to be given him after ten days of psychological torture.

That evening we were invited to dinner at the house of Aqqalu's cousin. He was even taller than Aqqalu, dashingly handsome, with straight white teeth and amber eyes. His name was Inunnguaq – we agreed to call him 'Inu' or it would surely have ended in disaster – and when we asked him what it meant he replied, 'Little sweet boy.' This is probably the least accurate name in human history, as Inu stood a good six foot six tall, and was built like a gridiron player. He was a superb host. His English was excellent, and he was confident with it too. He'd also invited along his uncle and a professional hunter who'd been tracking bear, seal and whale here for over 60 years. It's safe to say we learned more about Greenlandic culture in one evening than you'd get in a year living here.

Their house was extraordinary. With its weathered blue wood, and hangings of fishing nets and narwhal horns, it was almost too perfect. Set right down at the ice edge, you could sit on the balcony wrapped in your down jacket, nursing a beer,* and watch polar bears hunting out on the floes below you. Though it was ten at night, the sun was lighting up the mountains on the other side of the fjord.

---

* We were actually slightly thrown by the offer of a beer. Alcohol is a big issue in the Arctic, and we had deliberately been keeping our expedition whisky hidden for fear of offending anyone. Many of the communities here are dry by law, and you can get in trouble for bringing in booze. This is for good reason. Throughout the Arctic fruit and vegetables are scarce. In other parts of the world these have been distilled for alcohol to make wines and moonshines for millennia, but this far north people have not developed any physical or psychological resistance to the powers of the demon booze, and it has torn communities apart. Some peoples even lack alcohol dehydrogenases – the group of enzymes that break alcohol down. We were therefore nervous to accept a beer from Inu, but decided that on balance it would be OK. On the way home, though, we were approached by an inebriated local, who could barely stand and desperately wanted to embrace me. Inu was mortified. 'It is pay day,' he murmured. 'Some people here have no control.'

The family were fine company. Inu struggled to get a mouthful of food down, as we bombarded him with questions. After the inevitable trite queries – 'Do Greenlanders really have 100 words for snow?' (Answer: 'Kinda, but then so do you British, if you actually think about it'); 'Do you identify yourself as Greenlandic, or from this town, or Inuit?' (Answer: 'I am Greenlandic, from Ittoqqortoormiit, and my people have always been hunters') – the subject then turned to how this frozen world was changing.

'When I was a boy,' Inu told us, 'the ice was this thick [he mimed about a metre with his hands] and you could dog sled all the way to the mountains on the other side of the fjord. This year, it did not freeze at all. Even in the winter. Usually the season for Arctic char starts in June. This year, it started in February.'

'And what of the bears, Inu?'

'Most years we shoot four or six bears,' he said. 'This year, we have already shot 37.'

'Thirty-seven? But most people believe there will be fewer bears with climate change – why so many more?'

'There is no ice for them to hunt, so they come to shore. And they are starving, so they come to town looking for food.'

This fits with all my other experiences with Arctic peoples. Here, climate change is not an Excel spreadsheet with esoteric figures on it, or a meteorologist talking about fractions of a degree and feedback loops. Here it is their day-to-day reality. It is the cold (!) hard truth that their whole world is being turned upside down in the time it has taken a boy to become a man. And these lightning-fast changes are coming to a world more in touch with its heritage, with its past, than we could ever be.

'In the old times, everything had a soul. A stone has a soul, so don't move it unless you have to,' said Inu. 'If you take too much from the mother goddess of the sea, she will not give you any animals to hunt. And now times, if you throw trash to the sea, it is the same, you will have bad things come back to you. It is always this way.'

Perhaps it was his charm, his imposing form, his voice, but this simple phrase put the hackles up on the back of my neck. It was as if his whole people were speaking. As if we were being taught a timeless lesson. A lesson that I knew Western people, for all their technology, would be too late to learn.

As Head of Peril, Aldo asked if there had been any bear attacks recently.

'Yes,' replied Inu, 'there was one at Kap Tobin, where you were last night. The man went out to put fuel in his generator, and the bear ran for him.'

'What happened?' We were rapt.

'The bear ran for him, so he stood up and roared.'

'The bear roared?'

'No,' said Inu. 'The man roared. He shouts, "What are you doing here?" And then he punch the bear in the face.'

We were incredulous. 'He punched the bear in the face?'

'Yes, he punch it and then he come to town and tell us all. And then we shoot him.'

'You shot him?'

'The bear,' he clarified, 'not the hunter.'

Next morning, we rose ready to pack up our kayaks, only to find the fjord had changed once more. When we arrived in Itto-qqortoormiit, the western side of the bay had been open enough that we could have paddled through all the way to the huge area of open sea. Now it was thick ice you could pretty much have skidoo-ed across. It's something I've seen a dozen times and it never ceases to amaze me. The pack ice at this time of year can go from being thick enough to support a Sherman tank to being completely open water in a matter of hours.

'Would you Adam and Believe it?' I mused. 'It's proper thick.'

'You're proper thick,' said Aldo.

With an extra day to sort things out, we painstakingly packed up our kayaks. It's an art. A performance sea kayak has enough space for you to be totally self-sufficient for up to ten days, two weeks at a push. (More if you're prepared to 'grizz it out' and be a bit of a dirtbag.) It's only comfy, though, if you are thoroughly anal about your organization. Everything has to be in dry bags, no matter how watertight you think your boat might be. Sleeping bag and spare clothes have to be double-bagged. Every square inch of space has to be packed to avoid things sloshing about and ruining the boat's trim. You need to know where everything is, or it can take you an hour simply to find your toothbrush. Things that would be just annoying elsewhere become really important in the Arctic – the consequences of damp socks can be lost toes; a soggy sleeping bag could lead to hypothermia. Logistics, planning and running everything to a painstaking plan is the difference between comfort and calamity.

Another night darkened by a borrowed airline eyemask, and I was woken from a deep sleep, with literally no idea where or when I was. This was not helped by the fact that I was still wearing my eye mask, so it was pitch-black. Finally, my brain trundled into action and I realized what was happening. I ripped off the eye mask to blinding sunshine, and Rosie standing in front of me.

'It's clearing!' she said. 'We need to leave.'

'Amazing,' I murmured, probably not that convincingly. 'What time is it?'

'About 3.30,' she replied brightly.

'Jeez, Rosie, don't you ever sleep?'

A pre-dawn start is not nearly as satisfying when there is no dawn. You're staggering around with the fug of sleep tugging your deep reptilian brain, but don't get to satiate it with the sweetie of a sunrise. Given time, the midnight sun can really mess with your mind, sending you into a sleepless funk, where you barely register the passing of the days. On previous Arctic trips, we've battled on through long hard days of dragging sleds, only taking rest when our bodies are suddenly dog-tired. Inevitably, someone looks at their watch and notes, 'No wonder we're so knackered – it's 3am!'

The process of getting the kayaks down to the water was pretty over-dramatic for a humble paddle boat. The kayaks were loaded into the bucket of a JCB, which then trundled down the steaming cement paths towards the quay. Aldo and I, clad only in our tight paddling thermals, hung from either side of the cab as it rumbled through the snow alleyways. Both of us were in black from head to toe. We looked like a pair of disco ninjas. In my head 'Ride of the Valkyries' was blazing, and Robert Duvall was striding along the shoreline in a big Confederate army hat.

At the quay, it wasn't instantly obvious how things had changed – if at all. The pack ice was still in close to land, with just about every kind of ice visible from shore. We loaded up several small motorboats with the camera crew and all their equipment, and the four of us kayakers slid into the water. We opened up our shoulders, dipped our paddles, and finally, after years of dreaming, months of planning, and days of waiting . . . we were off. For about four or five strokes. Then it was all halt, as we hit walls of ice.

The progress was beyond painstaking. The surface was all frozen from the night before, black and crispy, kind of like paddling through the crunchy crust on a crème brûlée. Every paddle stroke skittered across the ice unless you samurai-sworded the blade down into it first. And this was the good bit. The thicker ice was as much as six metres deep, bobbing about in an ever-changing labyrinth. Several times we hopped out and dragged the boat over the surface of the ice, sinking into the slush on the surface of the floe. Other times we powered into smaller chunks of ice and tried to slide right over the top of them. We found that it was even possible to wedge yourself into a gap and push outwards, eventually driving even huge slabs on their way so you could fight your way through.

This ice was not icebergs. Icebergs are ancient, compressed monoliths, which come from glaciers. Snow falls on the ice cap or ice sheet way above, gets compressed over hundreds of years and flows down towards the sea. At the toe or snout of the glacier, this ice collapses or 'calves' into the water, and forms bergs. Icebergs can be the size of small countries (the biggest ever was the size of Jamaica) and are made of ice so dense it's like stone. What we were paddling through here, though, was pack ice – frozen sea – which had formed over the past winter. This can form into a solid shelf over the sea, which can extend for thousands of miles. Some chunks may survive for several years, but come early summer it generally starts to break up and melt. It is nowhere near as dense as bergs, so an icebreaker boat can power through it, and in fact we were doing a pretty good job of smashing it up with the bows of our kayaks.

The pack ice in Scoresby Sund usually stays solid till July. We were there in early June, and it was already well on its way out. The freeze and break-up of the Arctic pack ice has been fairly constant since the end of the last ice age about 11,700 years ago, and yet now, rather suddenly, the entire rhythm has changed . . . exactly as climate scientists have long said it would. The collapse of the sea ice does not have implications for global sea level rise – after all, the water tied up in it is itself sea. However, the shorter period of time that the Arctic is covered in frozen sea is enormously important. Firstly, because people and Arctic animals rely on it for their lives, seals to raise their young, narwhal to hunt around, polar bears to live their entire lives on. It is also the source of a major feedback loop in climate science. Just as you can see the reflective quality of sea and ice on the clouds above, the albedo effect of millions of square miles of ice serves as a powerful mirror, bouncing heat and UV rays back up out into space. Dark open sea does the opposite; it functions as a whopping great hot water bottle, soaking up all that heat, and leading to even greater increases in global temperatures.

At around 10am, after many hours battling and not getting very far, we could see a shimmering pond ahead of us, just beyond a crunchy channel. Smashing our way through, we glided out onto an inky lake amongst the white, the surface of the water without a single ripple, like smooth black velvet. It was exquisite. Arctic terns took off from the upper edge of ice floes, with white streamer tails open behind them, skit-skitter screaming in fury at us as we paddled beneath them.

Kilometres and miles here are meaningless. We probably only covered 15 kilometres that day, but it was a long hard day of slog, picking our way back and forward around the tide-borne ice. Late afternoon, we made land. A smoky fog over the hillsides revealed the spooky sight of looming shacks in the snow. This was Kap Hope, although a more appropriate moniker would be Kap Hopeless. It had been a hunting encampment, but had in recent years fallen into disuse, and was now a ghost town. Keeping weatherboard houses standing against the winds and snowdrifts requires occupation and constant vigilance. In the few years without a guiding human hand, the snows had taken back control. Opening the doors to some of the shacks revealed interiors packed with snow. Walls were caved in, roofs collapsed under the weight.

Kap Hope was a depressing place, dull as a gravel vajazzle with lumps of decomposing seal blubber lying in the muddy snow alongside chunks of asbestos and ripped-out electrical fittings. Though our instinct was just to hop back in the kayaks and find a place further down the coast where we could camp, a hut would obviate the need for bear watch. Finally we found one. It was no rustic hunter's cabin, however. Rusting appliances sat on mouldy mattresses and flea-infested sleeping bags. The ground was littered with old rifle shell cases, and it smelt of kerosene. I almost expected to see a human outline in white tape on the floor. The walls were scribbled in biro graffiti marking the score of previous hunting trips. One 2006 seal-hunting mission had vertical line marks for every seal kill. It looked exactly like the lines prisoners scribble on the wall in solitary confinement. Sarah, with her characteristic cheeriness, set about turning it into a home.

Beyond the disintegrating remnants of the town was their graveyard – extensive for a temporary settlement this small. The hard ground was littered with garish plastic flowers and unmarked driftwood crosses. A host of shallow graves bore testament to the harshness of life here on the edge of the world. Beyond us white Arctic hare gambolled on the scree slopes, and snow bunting chirruped cheerfully as they perched on the cross tops. It was a melancholy place, bringing to mind the words of Apsley Cherry-Garrard describing

Robert Falcon Scott's British Antarctic expedition in *The Worst Journey in the World*: 'And if the worst, or best, happens, and Death comes for you in the snow, he comes disguised as Sleep, and you greet him rather as a welcome friend than a gruesome foe.'

As the others trundled back to our *Trainspotting* squat, I stayed to think, cradling the rifle across my knees. My chilly white-blue fingers stroked the grey weathered driftwood of the crosses. They looked like the fingers of a corpse. *Really should have worn my gloves*, I mused.

The Arctic has been a fierce mistress to me over the years, and I feel differently about this environment than any other. On a still, blue-sky day, it feels as if it is scouring your soul. The emptiness, the ferocious potential of it. The sense of an environment that is totally oblivious to you and your presence. It's awesome. But that serene beauty can be snatched away in a second, replaced with howling gales and sideways sleet that cuts through you, making you feel simply irrelevant. I have been beaten here, and the very next day experienced highs too intense to put into words.

Wherever I travel in this high, brutal latitude I meet folk with meaty hands and grizzled features who seem to stare right through me. Men and women who embrace solitude, who talk in breathy tones, and set their jaws and brows to the boreal thunder. Who live here *because* and not in spite of the hardship. If pilgrims wander in the desert to find themselves, we come to the Arctic to lose ourselves.

As we penetrated further into the sound, the bergs increased in frequency and size. Some of them were enormous, and truly impressive, the size of towns. However, it was impossible to get any sense of scale unless you were right underneath them. One evening we sat watching an especially monstrous berg right out in the centre of the fjord. We all took turns guessing how big it was and how far away. The guesses went all the way from 50 metres high and four kilometres away, to 300 metres high and 20 kilometres away! With such disparity in perception, it seemed the only thing to do was to paddle in close to it. The main bulk of the berg was like the prow of a dreadnought or a mighty battleship, seeming to carve through the inky black waters. At one edge was a ceiling perhaps eight metres high, with vast icicles dripping down towards the water. It was the ice equivalent of our house of the dropping daggers cave in Mexico.

'Looks like something out of *Frozen*,' mused Aldo.

'Let it go, mate,' I said.

'Are we safe this close?' he asked.

'Ask Sarah,' I suggested. Sarah had paddled straight past us and was pootling along beneath the ceiling of daggers, so close she could reach out and run her hand down the ice.

'You wouldn't want to be under those if they fell,' said our Head of Safety/Peril.

'Yeah, I guess that would kinda hurt,' said Sarah, as if she were talking about a hangnail, rather than being impaled by a thousand ice harpoons.

'And is it true that 80 per cent of an iceberg is underwater?' Aldo asked.

'At least,' said Sarah. 'If this one rolled now, it would probably catapult us into orbit.'

'So this is just the tip of the iceberg?' said Aldo.

'Yes, mate.'

It was truly hypnotic, a sight of epic drama and wonder. With thousands of tonnes of ice in its structure, even a small chunk falling off could squish us like a summer gnat. As we watched, spindrifts of snow slid down its flanks with ominous swooshing sounds.

'I don't like it here,' said Peril. 'You two can mess about here if you want, I'm off for some tea.'

(Aldo and Sarah had become fast friends, which is weird as they didn't seem to understand a word each other said. I mean, none of us generally understood a word Aldo was saying, but for a Canadian hearing a Glaswegian in full flow, it must be like trying to decipher Swahili.)

'I'd like to paddle through here in the fall,' Sarah said.

'Autumn,' Aldo corrected her.

'I reckon when more of the ice has gone, this would be the most stunning route,' she continued, ignoring him.

'That's route,' he replied. 'As in, rhymes with "shoot", not "shout".'

'What's he saying?' she asked me.

'I really have no idea.'

'And when I make pasta I use Ore-GAH-no, not Or-EG-ano, and BA-sil, not BAY-sel.' He is into his swing now. 'Learn how to speak English.'

'Glass houses and stones,' Sarah tuts.

'Seriously, you Americans murder the Queen's English.'

Sarah is of course Canadian and is not biting. 'Did you learn that in the army?'

Boom! I watch Aldo seethe with fury. I have taught her well.

That day was to be the trickiest of the whole paddling section. We'd hoped to stick in close to the coast, but found the pack ice had been swept in by easterly winds, and was hugging the shore in impenetrable rafts with jumbled blocks and shattered pillars. We were not ready to give up so soon, though, so decided to try to paddle through the floes and out to the open sea beyond.* Progress was painfully slow, and by the middle of the day we had travelled no more than ten kilometres from the last night's camp. What was worse, we were way off shore, and there was no getting round the fact that the ice was closing in around us. It was like a labyrinth that is always growing new corridors and shutting off old ones behind you.

It was such a strange energy here. Within the ice you are in a world of impossible tranquillity. It's quiet, except for the occasional

---

* While this initially proved easy for the kayaks, it was not so for the filming boats, which struggled to keep pace with us. It didn't help that the two boat-men were entirely different in their attitude to the ice; one clearly experienced and confident, the other much more nervous. Our confident driver would ram his tiny plastic boat into the ice, then power up his outboard motor and simply push the floes aside. His colleague would falter and then get trapped as the floes were driven back again by the wind.

In the 21st century, if you want to be an explorer, you want to be a cave diver. There are untold miles of cave as spectacular as this that have never been explored before.

The aerial view showed us the scale of promise in Yucatan, forests filled with potential.

Getting down into the abyss affectionately nicknamed 'Backshall's Backdoor'!

This jaguar mural was emblazoned in charcoal over a thousand years ago
– the location took many hours of caving to get to.

This was one of the most decorated caves I've ever seen, with Mayan
artefacts, and a shimmering undived lake at its furthest reach.

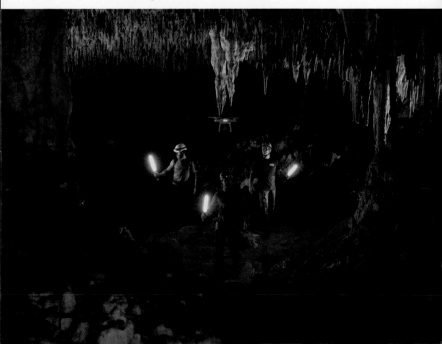

Picking our way
through the leads
between the pack ice.

First thing in the morning; it's glassy calm and we're breaking a crispy crust of ice on the surface with every paddle stroke.

As climate change leads to ice breaking up earlier and forming later, more desperate bears will be brought into uncomfortable contact with people.

Dragging our gear up onto a vast Greenland glacier and towards a range of unclimbed peaks.

By using skins on the underside of our skis, we could get up all but the steepest slopes.

2.30am, standing on the summit of the mountain I named after my first-born son; an emotional and exhausted moment.

Icebergs are some of the most transfixing and terrifying of natural structures, unstable and liable to fracture or roll at any moment. We kept our time near them to a minimum.

Looking down in Jebel Shams, the Grand Canyon of Arabia.

Despite its diminutive size, the saw-scaled viper is one of the few snakes that are a significant danger to human health.

Majlis al Jinn is a true forgotten natural wonder; at the right time of day beams of light create a breathtaking spectacle.

shrieks of bothered Arctic terns. But calm and peaceful as it may seem, it is a dynamic environment that never ceases moving, and sometimes in calamitous ways. The inexorable shifting of millions of tonnes of frozen water can carve valleys, create mountains, transport apartment-sized boulders for many miles . . . and splinter a boat as if it were made of Ryvita.

The Arctic is a place that puts your moods beyond your own control. On a blue-sky day where the sun sparkles fluid gold over the sound, my spirits soar to an intensity I cannot replicate anywhere else. When the winds howl so they cut through you, and the greyness of the sky turns the whole world into a blank, bleak concrete drone, my mood plunges. And when my mood is low, sometimes it seems my negative energy manifests in everything around me.

Putting up my tent I struggled against the wind and, pushing a pole into place, managed to spear the end of the pole clean through my nice warm down jacket. There followed an explosion of feathers like an Eider duck smashed on the wing by a gyrfalcon, and my only warm jacket was punctured. Exasperated, I got inside my tent and went to zip up my sleeping bag. The zip broke. I'd had that sleeping bag since 2004 and it had become my lucky blankie, my solace. Like Linus from *Peanuts*, I clasped it to myself for comfort when I was feeling low. Like now. And now even my blankie had let me down. Out at sea I had reached out to grab my water bottle and dropped my only dry hat into the water. My head wouldn't be warm again till the end of the trip. On such things pleasure and misery balance. Coming back to the tent again, I sat on my inflatable sleeping mat to take my boots off. There was a bang so loud that Aldo leapt from his tent with gun ready.

'What the hell was that?'

'Just my sleeping mat,' I said glumly.

'You've popped it?' He chuckled with sadistic glee. 'Sucks to be you!'

Aldo is always delighted to see the results of my 'bad admin'. Being a military man, his tent is organized with near manic precision. His wash kit is placed alongside his immaculate sleep set-up as if it is waiting for use in an expensive hipster barber's, everything aligned on a spotless flannel. I swear he carries a ruler with him to ensure everything is the exact same distance apart. His boots sit outside, alongside ice axe, always sharpened, and rifle, always oiled and spotless. His favourite thing to do is to walk past my tent, peer inside and shout out, 'Quick, Stevie B, look at your tent, you've been burgled!'

My favourite thing to do is to go into his tent while he's not looking, turn his toothbrush round 90 degrees, put his beard brush into his boot and my rubbish in his sleeping bag. It sends him berserk. I introduced Rosie and Sarah to this game. You knew when there had been an untidying, because there would suddenly be a distant roar like an angry bison: 'Aaaagggghhhhh! I swear I'm going to take you guys out in your sleep – they'll never find your bodies!'

Pretty soon, he was spending the whole day following us around, trying to catch us in the act.

'It's not very smart to always be trying to screw over the guy who has the gun.' He had a point.

'We can't help it, Aldo, your OCD is just too tempting.'

'Admin is not a place in China,' he replied. This is his favourite phrase.

Despite my teasing, this is actually one of the reasons Aldo and I make a good team. If I didn't have Aldo, always organized, prepared, covering all the bases and getting all the 'admin' in good order, I'd have done myself a mischief years ago. And if he didn't have me constantly bending and twisting the rules and the playbook up into knots, we'd never do anything cool! Not many ex-military men could put up with my scattergun approach to expeditions, or my cavalier attitude to organization, common sense and personal freshness. Most would have smothered me in my sleep on week one. We have never argued or got cross with one other. He has a genuinely relaxed nature, a willingness to give anything a go, and good judgement as to what is important and what isn't. Aldo will always be the one up first and to bed last, already making coffee when you're still wiping sleep from your eyes, burning off the most sweat, sleeping in the crappiest camp spot. There is also great comfort to be had from knowing that whatever hell you put your partner through, it cannot be as bad as war. Or providing deep protection for 'narcos' working with the Colombian cartels. I'd love to say he was the Watson to my Holmes, the Irvine to my Mallory. Sadly, it is more like he is the Blackadder to my Prince George.

The further we penetrate into the fjord, the more the scenery evolves. While the southern side soars with vast mountains, with gargantuan glaciers tumbling down between them, our shore is undulating brown and white hills. It's bleak, verging on grim. However, day-by-day the snow is receding and the vegetation is leaping into bloom. We spot Arctic cottongrass and tiny purple saxifrage flowers, as well as yellow Arctic poppies and raging violet fireweed. Straggly dead twigs on the ground suddenly produce shoots and catkins – these are dwarf willows and birches that never get any more than ankle height, yet could be many decades old. They are food for a bamboozlement of tundra wildlife, one of which we see more and more on the hilltops as we progress. They look like dark shifting boulders. Musk oxen stick to the hilltops at this time of year as the biting winds blow them clear of snow before the valleys. Named for the strong smell of musk which they stain their guard hairs with, these walking carpets are some of the most bizarre and brilliant beasts of this winter world. At the rut, the males will charge towards each other at 60 kilometres an hour, colliding heads with a finality that judders your neck, even when they're a mile away.

'They look like Arctic versions of a Shetland pony,' says Aldo.

'You're pony,' I reply.

On the ground near camp, I find some of their old droppings. They look just like Maltesers. Then I am caught by a sudden miraculous realization – we have a pack of Maltesers in the snack store. I go and rummage through the stash and find them, then nonchalantly sit down in the sun near Aldo and start conspicuously eating them. Aldo is drawn to chocolate like a moth to a lighthouse.

'Giz one,' he says.

'Open up,' I motion. He opens his mouth wide, I aim, and toss a Malteser across. He tries to catch it in his mouth and just misses. 'Again,' I say, and toss another. He catches it and devours it. I throw one more. He swallows it. The trap is set. I take the musk oxen dropping and aim as if going for a triple twenty with my life's savings riding on it. The poo bounces off one of his teeth and into his lap.

'You'll have to get up earlier than that to catch me out,' he says.

I check my boots for booby traps every day for the rest of the trip, but it was worth it. Almost.

Someone once said, 'There's no such thing as bad weather, just bad clothing.' That person never tried to paddle a kayak through shifting Arctic pack ice in a howling gale. It may be possible to be comfortable in the most severe conditions, but that is only while your effort levels are continual and consistent. When your activity levels vary in intensity it is impossible. On the first day it was mostly sunny and I cooked. On the second day I stripped out a thermal layer under my drysuit and went with a much thinner base layer. This was great for the first three hours, then we hit a freezing fog and I felt as if my nipples were being set about with a scouring pad. After ending that day borderline blue, I put on a fleecy bodice underlayer. The pack ice suddenly opened up into open sea and thus hours of hard paddling without stopping. I eventually had to pull into shore and re-dress because I was soaked with sweat. We then hit pack ice again and had to slow down to pick our way through it. I felt as if I'd been dropped into a fridge freezer in my Speedos.

While mammals had been in short supply since we began paddling, the birdlife was exquisite, with a distinct air of manic activity. Flocks of eider and king eider stretched to the horizon, little auks ducked and dived, storm petrels danced on the water with their dainty feet tapping the surface. The most wonderful sound was the haunting wail of the Great Northern diver, known in America as loons, because their mournful yodelling sounds like far-off escaped lunatics. There were busy turnstones, also known as brick flickers for their ADHD scampering, tossing over every pebble in their path. And red phalaropes, another bird possessed, whirled round and round the shallows in tight frenzied circles, whipping up aquatic invertebrates to feast upon. Every one of these winged weirdos was in overdrive, frantically feasting on the plenty afforded during the few short months of the Arctic melt.

It's a very clear reminder that the Arctic is not barren. Far from it. It may well feel inhospitable to us, but that's because we're a mammal that evolved in hot and temperate climates. Arctic terns migrate across the entire planet to be here at this time of year for a reason. They could stop off anywhere – they literally have the whole world to pick from – and they come here. If you've got thick feathers or blubber and dense hair, the weather here probably feels balmy. The seas this time of year are teeming with fish and other marine organisms. There is a relative paucity of terrestrial predators. The animals that live here, or migrate here, do not endure hardship in being here. They have evolved into physiological forms that find this climate comfortable, with temperature, conditions and rhythm of the seasons that have been predictable for several million years . . . until now.

Everything here exists on a knife-edge, and when things change, they suffer. If the ice here was to melt over geological time – a thousand years or so – then polar bears would adapt. Now that it is changing over decades, they have no chance.

It was day three of paddling when everything changed. We were camped out on a broad flat beach with hills behind. Bergy bits and pack ice bobbed around at the seashore, and dunlin and sanderling picked around for scraps on the narrow strip of sand. It was around 8am and we were packing up to leave, when someone said: 'Bear!'

We looked down the beach and there it was. Ambling through a searing heat haze was the great white bear himself. He was a good mile away, but this was the encounter we had all been waiting for. We chatted enthusiastically about what an exciting experience it was for us all, how we were truly in the realm of the ice bear and about how his attitude was relaxed. How we were upwind of him, (he was downwind of us) so he would have smelled us hours ago, was not interested in us, and would probably avoid our camp. But then he just kept coming. Steadily, minute by minute, with that slow lolloping gait, he ate up the metres. Every now and then he'd lift his nose to the air and sample our scent.

As he got closer, his attitude changed. He was no longer ignoring us. Not only was he smelling, but he was giving us huge exaggerated yawns. These can mean many things, but it's always very deliberate. Sometimes it can be 'displacement behaviour', the equivalent of when the team and I are terrified on a cliff and one of us tells a bad joke. It's classically when an animal is torn between two drives, on one hand the desire to approach us, on the other the fear of what that might lead to; so it does something else totally unrelated. It is a way of alleviating stress, through an alternative behaviour. On other occasions these yawns can be a threat, and a precursor to a charge. It was impossible to interpret the signals. However, there was no doubt that he was getting closer and closer. And showing no signs of slowing down. Eventually he was no more than 30 metres away from us, and we had to start showing him who was boss. I got up in his face, yelling, shouting, clapping my hands and moving towards him. This is usually enough to deter most bears, but it didn't work. Sarah charged up one of the flares and fired it. The burning red pixie popped and bounced around his feet. He turned and ran. Perfect! He'd make a run for it now.

Except he didn't. He turned and came again. I could see now that he was a younger male – perhaps third or fourth year – with a big lateral scar across his face. Possibly from a fight with a bigger male, or even as his mother drove him away to force him to disperse.

These males can be the most dangerous. They are inexperienced, and at this time of year as the ice recedes, they can find themselves hungry and hurting and coming into contact with people. Like us.

Out of the corner of my eye, I saw Aldo picking up the rifle and cocking it, putting a round into the chamber ready to fire. 'Please, please don't use that unless you absolutely have to,' I begged him. He nodded.

One of the boatmen fired a shot over the bear's head. The bear turned his head away, but for no more than a few seconds. Then another flare fired, which he completely ignored. This was turning into a nightmare scenario. An inquisitive bear will turn and flee at signs of resistance. One that keeps coming is trouble. And when you have used up all your non-lethal methods, you have no choice but to shoot to kill. This was what scared me. I was not frightened for myself, even though I was standing only metres away from an animal that could tear me to shreds. I had a Royal Marines sniper standing at my shoulder with a rifle, and Arctic Sarah with a 12-gauge shotgun right next to him. If the bear had charged, it would not have reached me. And that would basically end everything.

To kill a rare polar bear, up here, in this wilderness . . . that would make every altruistic action of my life meaningless. It would make a mockery of all the years I've spent trying to convince people that we should not be scared of wild animals, and make my legacy nothing but that of the imperial hunter, taming the environment with bullet and brute force. I saw myself standing over the cooling corpse of the great white bear. That could not happen.

I ran forward, shouting and screaming, and stooped to pick up stones to throw at the bear. It appeared to be a cruel act, I know – but I was throwing stones in order to save his life. I needed to prevent my friends from having to pull the trigger and drop him to the sand.

Another shot, this time into the ground in front of him. He ignored it. He was coming again, too close now. A stone pinged in front of his nose and he didn't flinch. Then 'thud' – I hurled a good shot, and one hit him on the flank. He pulled away this time, as if stung by a horsefly. That was unexpected. We backed up our advantage, charging at him, yelling. This time he broke into a brief trot, looking over his shoulder as he went. The magnificent bear stopped again, as if making up his mind. Were we worth the effort? Finally, mind made up, he looked to the sea, wandered down below the snow shoulder and disappeared. This was perhaps the most frightening thing of all. Our half-tonne bear just vanished into thin air, behind what we all knew was a thigh-high ridge of snow.

The crew all watched in tangible relief as our bear slid into the water, and started to swim away from us. His white head bobbed along the surface, its menacing potential lurking beneath, just like the bergs around it. Finally, he pulled himself up onto a piece of floe ice, and shook like an oversized ivory Labrador before rolling around on the ice like a puppy rolling in fox poo . . . except this puppy weighed as much as a Friesian cow, and could kill a whale with a cuff of its paw.

Climate change and its after-effects are not the only threats looming over the King of the Arctic. Respect and awe have not prevented as many as 300,000 polar bears being killed from the beginning of commercial hunting in the 1700s through to 1973, when the 'Agreement on the Conservation of Polar Bears' was put in place.

There are still around 1,100 bears allocated annually as 'harvests' for subsistence hunting, practised by native Arctic peoples. The Inupiat have similar annual quotas to hunt the unicorn-horned narwhal, white beluga whales and 100-tonne bowhead whales. Though this is classified as a traditional subsistence industry, hunters don't use hand-thrown harpoons and rowing boats as their ancestors did, but often use plane and helicopter spotters, then shoot whales from motorboats using explosive harpoons and automatic weapons.

In the Nunavut region of Arctic Canada (where I've done several expeditions), local people enter a lottery to gain 'tags' to allow them to shoot a polar bear, worth as much as $10,000. However, those tags are being auctioned off to foreign hunters. Just type in 'polar bear hunt' into Google, and you'll find outfitters that, for a hefty fee, can hook you up to shoot yourself a bear.* The pattern

---

* This constitutes trophy hunting, which is clearly in breach of the 1973 accord. In their paper, 'The Economics of Polar Bear Hunting in Canada', the authors

seems to be the same as that found in African trophy hunting. The wealthy client will fly out to an Arctic location, be put up in comparative luxury, guided out onto the ice by knowledgeable locals and given the chance to shoot a bear at close range, before the resultant head or rug is flown back to the tourist hunter's home nation. While there is undeniably some money filtered back into the native communities by this kind of enterprise, most of it goes to big game safari companies run from metropolitan cities. For example, it's estimated that trophy hunting in Nunavut represents only 0.1 per cent of the country's GDP. And as for the trophy hunters themselves – at least African big game can occasionally escape! On the tundra and pack ice there are no trees or hills to hide behind, and polar bears are so well-insulated that they cannot travel far fast as they dangerously overheat. So once one has been located, it has simply no chance; it's just a matter of time.

All of this is done above board and is legal. In Russia, where hunting is not allowed, many hundreds of bears are poached illegally, which of course happens across the Arctic. On a filming trip to the north coast of Alaska, we chose one of the two local guides to take us out looking for polar bears. The next day we met the

---

explain that there was no tradition of trophy hunting in Canada's Arctic regions until the 1980s, when the territorial governments encouraged the construction of the industry to bring money to people in far-flung outposts of the nation. They report that trophy hunting has since increased twentyfold there, with indigenous people selling off their allocated quota of polar bears to wealthy outsiders, mostly from the US, Europe, China and Russia.

other – less reputable – guide out on the ice with his 'bear-hunting dog' and a high-powered rifle. He'd not shot any polar bears yet that day, but had made do with a brace of Arctic fox instead, which the night before had been playing round my ankles like puppy dogs. We spotted a distant wolverine strolling across the pack ice, and were informed none had been seen in the area in at least 14 years. Minutes later, two 100mph snowmobiles screamed out across the ice, their occupants toting assault rifles. That evening it was lining our inn-owner's jacket.

So what can be done? Upgrading Polar Bears to CITES I protection status would mean foreign hunters would not be allowed to take any part of their trophy home with them. Somehow, a photo album and a home movie is not quite as strong a centrepiece as a taxidermied bear, and might put some off hunting. Language is important. Politicians can't talk of 'harvests' – this isn't wheat, they're rare and precious animals. The US government's recent decision to repeal legislation on bears, however, allowing hunting in national wildlife refuges of cubs, mothers and babies in maternity dens, including shooting them from helicopters and planes, was a major step backwards.

Bears wander over the whole Arctic, and do not belong to any one region. If there genuinely is a surplus of polar bears, which allows for sustainable hunting, this needs to be ascertained through peer-reviewed science, which is not undertaken by those who have vested interests. Few of us will ever see the wonder of the wild white top of the world, but that doesn't mean it can be out of sight and out of mind. We did that with our oceans, and are now start-ing to reap the whirlwind. And while the IUCN* may list the chief risk to polar bears as being the decline of their icy home due to climate change, hunting them for wall-mounted trophies is cer-tainly not going to help.

The day of our close encounter with the polar bear, the winds were with us and we made easy distance, putting the miles behind us and pushing further up the fjord. Late afternoon, Rosie pulled up alongside us in the filming boat.

'What's your plan, Stevo?' she asked.

'Well, while the conditions are so perfect we should make some ground,' I suggested.

'OK . . .' She sounded unsure. 'The boys have quite a lot of work to do, though.'

'No worries,' I said. 'Let's make camp as soon as we find some-where decent.'

We pulled in no more than an hour later, and went through the painstaking process of dragging boats and equipment inland and uphill, before making meagre flat ground into viable sleeping spots. My bear watch was three to four in the morning, yet it was as light as midday. I woke up to a brutal, biting cold, and

---

* IUCN stands for The International Union for Conservation of Nature; CITES stands for The Convention on International Trade in Endangered Species of Wild Fauna and Flora.

an eerie silence. Bear watch is definitely the time you feel the Arctic solitude the most. If others are awake, inevitably the stoves are blaring and the little generator is rumbling, so you never get the full hit of northern silence. Now it hit me like a blast. The quiet was overpowering and haunting. Much like in desert nights, the silence is so quiet that it's loud. It seems to roar in your ears and howl at you like a voiceless banshee. Few of us managed to sleep both sides of the shift, so it tended to halve the amount of sleep we were getting as a group. People started to look tired and to act grumpy.

About halfway through that evening's bear watch, all of a sudden the wind swung right round. It had been blowing from behind us and turning our bodies into sails. Now, and for the next few days, we would have to battle into a headwind, the true enemy of any kayaker. I had hours and hours of shoulder-wrecking paddle strokes to look forward to, as I cursed myself for not continuing while conditions were good.

The next few days were more what you might expect from an Arctic sea kayaking expedition: an inescapable and ceaseless chill, bleak black skies, ugly green seas and a general feeling of being in an environment that hates you. The wind blew so incessantly in our faces that we couldn't stop even to sip a drink or put on an extra hat. If you did, you'd instantly start going backwards and end up where you started. Every mile was hard won, and as boring-looking as the mile before.

Finally, we hit the remainder of the pack ice after eight days of hard paddling. Whereas all the pack ice we'd seen till now was fractured and marked with features, a jumbled mess that had been cracked up and then been borne about on the wind, this consolidated pack ice was uniform in appearance. It was a blanket of white extending out to the horizon, with no hills or even undulations, just plain frozen sea. At the edge of the ice, the water had the same quality it had in the halocline in our Mexico caves. The texture was greasy and oily where the melting freshwater encountered the salty seawater.

As dull as it might first seem, it's obvious why this place is so important. We saw fleeting glimpses of two seals on the entire paddle out here. At the ice edge, however, seals are spaced out every few hundred metres. There are dozens of them. These were the young of harp, ringed and bearded seals, left here by their mothers after being born three months ago. They can swim, but don't have enough fat insulation to be in the water for long. They are a nice wee snack for a passing polar bear, who'd scoff one at a single sitting. However, they are only *here*. And if the pack ice breaks up too early, they will not survive. And the bears will have nothing to feed on. It's all such a delicate balance.

Once we'd reached the pack ice, my plan had always been to haul our kayaks across it until we reached open water again. However, the pack ice was no more than half a foot thick, less in places, and would certainly not hold our weight, at least not the entire way. Also we were still 80 kilometres from Bear Island, where the fjord officially branches off into its subsidiaries, and even on good ice that would take a week. Satellite images gave us no real proof that there was any open water beyond the white. We could have broken our backs doing it, put ourselves out there, committed and exposed, and then never found any more water to continue our journey. For now, the kayaks were done.

This meant that the only way on was to hop over this section of pack ice by helicopter, something we were all very loath to do. It taints the purity of any expedition; we all wanted to be travelling under our own steam, by muscle power alone. The irony of being beaten by ice when we were talking so much about climate change was not lost on me. But we had no choice.

Our journey took us up to the Doughard Yansen glacier, at the head of the fjord. Our helicopter flew over the Bear Islands, where we had most hoped to be able to kayak. All of us were crossing our fingers that the sight would be underwhelming, and therefore we could all be at peace with how the expedition had panned out. Sadly, this didn't happen. The islands are black peaks sealed into a snooker-table flat blanket of pack ice, with glowering mirror-faced mountains all around them. Before the islands, 'Iceberg Alley' has thousands of giant bergs frozen into the white, as if in stasis, trapped here before they can continue their journey down and out to sea. They look like snowy mountain tops poking up out of a cloud inversion. I almost could not bear to look, it was so intoxicatingly beautiful. *One day I'll come back with Helen and paddle through here*, I thought.

From the Bear Islands, our heli penetrated up into the north-west arm of the fjord, and into one of the most incredible landscapes on earth. Monstrous mountains cascaded down into the frozen fjord, with almost infinite new climbing routes on offer. Glaciers too immense to contemplate tumbled down from the Greenland icecap – the largest on the planet outside of Antarctica. The king of those, though, was at the terminus of the fjord. It's one of the most productive glacier on earth, letting rip with 18 cubic kilometres of icebergs every single summer. That's over 17,000 Empire State buildings worth of ice. The toe or snout of the glacier was a jumble sale of skyscrapers of blue and white, and disgorged bergs the size of castles, dwarfed by others the size of mountains. It looked like Krypton, the planet where Superman and General Zod were born. It was without question the most impressive glacier I'd ever seen.

A cliff face hung over our campsite like a vast breaking wave, hundreds of metres of sandy granite cut through with dark basalt intrusions, like the chocolate filling in the wafer biscuits I used to get in my school lunchbox. The curved cliff face amplified an angry clucking call: the screams of a gyrfalcon chick calling out to its errant parents to bring it a snow bunting snack. Gyrfalcons are the largest falcon in the world, and the most beautiful. Arabian falconers will pay hundreds of thousands of pounds for a pair of these Greenlandic white morph gyrs. It's a miracle they exist in the wild at all, their survival probably only made possible by this propensity for nesting in impossible crags, where even the most hardened poacher cannot steal them.

Our campsite was high on a ridgeline, in amongst mountain

avens, with butter yellow sepals and whispery white petals, and mountain sorrel, its red-tinged leaves rich in vitamin C, fungus like strange flowers, as well as dark blue bellflower. The bright red Eurocopter sitting there on the ridgeline was the most incongruous of centrepieces to the camp, but added to the certainty that no one had ever camped there before. The Austrian pilot Christian was delighted to spend the night there with us, freed from the boredom of ferrying fuel and people out of Constable Point and thrown into the bizarre world of a Backshall exped, camping on a mountainside with a view to rival any on the planet.

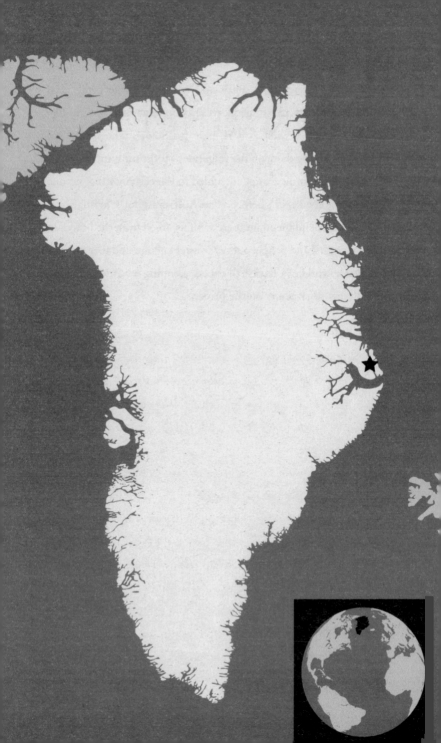

4. FROZEN PEAK

GREENLAND

Gazing out to the horizon at this height you can see hundreds of miles, yet we know we are the only humans in the entire landscape. It's 2am, but the thin sun is still half-heartedly trying to shine through stringy clouds. Our reverie is shattered by a creaking sound like an overburdened teak table – a crack then a rumble. 'Rockfall!' We thrust our faces into the crumbly cliff face, cheeks pressed to bone-chilling rock. There follows a full Doppler effect: distant growling, before the rocks fizz past us, thundering in our ears, then thudding into the snow below and rolling off, gathering speed and snow. I look around. Everyone is OK. Normally this would be a moment for the team to stop and breathe heavy, sidelong glances telling of a close call and near miss. But not on this peak. On this climb rockfall is the norm. Every single footstep and handhold is an avalanche waiting to happen. Kicking heavy snow from my crampons, I swing my ice axe into the crispy snow and move on.

While the fjord is the most notable feature carved into the east coast, it's also home to the Stauning Alps, a mountain range the scale of the European Alps, though with its highest peaks not quite reaching 3,000 metres above sea level. The *average* annual temperature here is minus 16 degrees centigrade, but as we were to be here over the summer solstice, temperatures could well reach a balmy minus two. Some of the peaks here were climbed in the early 1900s, and any ascents have been well mapped and logged with the Danish Polar Centre, Royal Geographical Society and British Mountaineering Council. So we came armed with as good a map as exists, and a colour code. Peaks coloured in red had been climbed before. Those that were green had not. The north away from Scoresby Sund is characterized by firm granite, and had more reds than greens. The southeast, though, had several valleys where literally every peak had no recorded ascents. As we flew round the summits by helicopter, we compared the peaks in real life with what we could see on our map, as we planned, plotted and fantasized.

This next section of our journey was to begin in Schubert Dal, a river valley to the east of the glacier. We knew little of the valley, other than what we could discern on satellite imagery. Certainly it was at least five kilometres across, and not steep enough to contain much challenging white water.

We landed on the eastern bank. The valley was scenic in a vague, bleak fashion, with snowy mountains on the far bank, and steep mud and scree hills on our bank. As I stood putting up my tent, musk oxen watched me suspiciously through a raging heat haze from a ridgeline no more than 50 metres away. We had planned to trek from here through the hills to another valley, which is a hotspot for breeding birds, and hence also predators like the Arctic fox. However, on arriving at river camp, we started to revise our plans.

The melt was far more advanced here than anywhere else we had been, and the ground was a soggy, boggy mess. We would need to do the 25 kilometres walk several times in each direction in order to get our big camera, tripod and lenses into a valley camp. In addition, long-range weather forecasts showed we would have at least five days of high pressure (thus blue skies) followed by uncertainty and probably rain. It seemed the best option was to change our plans and head straight for the climax of our expedition – first ascents of summits in the Stauning Alps, and on the other side of the Schubert Dal River. It turned out to be a decision that saved our entire expedition.

River camp was also where we said a sad farewell to Sarah. She had been a beacon of sunshine throughout the entire time she'd been with us – selfless, cheerful, fun and effortlessly capable. She would be much missed, and we all rather wished we'd had the foresight to include her on the mountaineering section of the trip. Though it wasn't her speciality, there was no doubt she'd have taken it in her stride.

Her place, though, was being taken by two mountain guides, Libby and Tamsin. Tamsin was chirpy, smiley, strong as an ox and infinitely competent, with an accent that roamed all over the British Isles, drifting into a Scottish brogue, while belying her current domicile of the Alps. She had the most impressive Panda eyes from her sunglasses I've ever seen. Libby was tiny, with a slender climber's physique, probably half my bodyweight, but with a command and carriage that told of huge depth of experience. They slotted effortlessly into the team, and it was soon hard to imagine our little family without them.

As we were now well away from the coast, it was also a time to put bear watch aside. We were pretty confident a bear would only rarely wander through this valley, and so instead decided to go with bear fences around the tent. These vary in complexity from electrified cow fence-type scenarios to just wires with your pots and pans on them, so they jangle when a bear walks into them. We had a mid-range option, with fishing wire through broom handles attached to rape alarms. Far from helping us to sleep in safety, it resulted in us being woken countless times as the broom handles toppled over and set off the high-pitched wailing. The next night we decided they were less than useless and went back to bear watch, while I grumbled that no bear would ever come up this valley.

Now that we had abandoned our original plans, we needed to cross the river. It didn't look like much to the casual observer, but we'd crossed more than our fair share of Arctic rivers, and were all dreading it. It didn't look as if it would be massively deep anywhere, or treacherously fast-flowing. It was, however, very wide, with braided channels of brown water interspersed with sand and mud banks of uncertain quality. As the water would have been glacial ice no more than hours beforehand, it would be punishingly cold, and may well repel our efforts completely. We all sat for many hours discussing different strategies, clothing and footwear possibilities, before coming to a consensus. It was going to suck, and we would just have to lump it.

The next morning we packed ready for the long slog across the river and on to our next basecamp. We estimated this was no more than 20 kilometres away, but the terrain made that figure almost meaningless. Aldo and I had also managed to turn the gentle stroll into a death march by dint of some of the worse 'admin' imaginable. When we'd been dropped off by the helicopter the previous day, we'd had to hurriedly pack away stuff we wouldn't need for the next section of the expedition so we could send it back to Constable Point in the whirlybird. Both of us, however, had been distracted during the rush and were perhaps a little careless about what we'd sacrificed. Now, in the cold light of the endless day, we each found ourselves with huge rucksacks, filled with stuff we wouldn't need.

I had all my research notes, two books on Arctic wildlife, a coffee maker with no coffee and four different pairs of boots. Aldo had an entire menswear department of retro-climbing apparel, hoping it would make him look like Ernest Shackleton. When my pack was full to bursting, Keith came over with all his gear. I had offered to take it, to leave him load-free for filming. The only solution was to tie things all over the outside of my pack, so I was jangling like a 12-year-old who had borrowed his uncle's rucksack for his first Duke of Edinburgh expedition. Aldo, if anything, looked worse.

'God, can you imagine the comments we're going to get on Twitter for these packs?' he whinged.

'Shut your moaning,' I chided. 'At least I haven't packed a beard brush.'

'You need testosterone to grow a beard like this,' he said. 'You'll never need a beard brush.'

'All dwarves say that, Gimli son of Gloin,' I replied. 'Even the women.'

'Seriously, I look like a total amateur, all the guys in the Marines are going to have a field day.'

'I'm guessing by the end of the day you won't care what anyone else thinks,' I noted, trying and failing to get my oversized monstrosity of a pack onto my back. It was a bit like trying to get a saddle onto a moving elephant.

Before we started our walk, I strolled round the mound we were using to obscure our ablutions, toilet roll in hand. And walked straight into a musk oxen, distractedly chewing on dwarf birch. He stopped mid-mouthful to regard me balefully, unsure whether to run away, ignore me completely or charge. It struck me that this would be a most undignified way to go, punted into the troposphere by a grumpy midget bison with my trousers round my ankles. However, the whole team was waiting, so I got on with it. It was very difficult to ignore his watchful gaze, no more than 15 metres away while I got down to business.

'I say, old bean, you wouldn't mind looking away, would you? It does somewhat tend to put me off my stride, if you know what I mean.'

He didn't avert his eyes, but soon got back to the serious

business of chewing, deciding that I must be harmless. As I wandered back to camp, the big bull strolled the other way, and into view of everyone else.

'That's amazing,' said Aldo. 'He just isn't interested in us at all, is he?'

'I think he's kind of keeping us under close supervision, but just making a big show of not being bothered.'

It was certainly a great sign. That the big male wasn't stamping off into the distance meant humans did not penetrate this far into the interior. It was a sure indication of the genuine remoteness of our location, the wildness and loneliness of our river camp, which was just the start of our adventure.

'Kind of reminds me of you,' I mused. 'Ginger, angry and shorter than he looks in photos.'

'He has my haircut too,' Aldo agreed. 'Or Tina Turner's haircut from the eighties.'

If the nonchalant wandering of musk oxen into our camp was something of a surprise, after no more than 15 minutes of walking we hit something even more incredible. The tundra was a fairly uniform brown colour; clumps of flowers as yet isolated, mere splodges of colour amongst the dull. But ahead of us was a small bright green mound, which from a distance looked like an exhibit at the Chelsea Flower Show.

As we approached, we tiptoed over the vertebrae of an ancient spine, moulded into the tundra and overgrown with lichen and moss. The centrepiece, though, was the skull of the musk oxen.

The boss and horns were unmistakable; the bleached white snout of the dead beast protruded from the soil as if it were some kind of zombie ox, about to erupt from its grave. The nitrates and phosphates leaking out of its bones were a potion fertilizing the soil around the skull, bringing it to life. Green shoots sprouted around the head, mimicking the outrageous hairstyle of the living animal. In between the grasses bloomed miniature flowers of purple saxifrage – the first signs of spring up here – as well as alpine bistort, with long leaves, elongated red stems and white flowers, and snow buttercups, creeping cinquefoil and Arctic dwarf willows. There must have been 15 species all erupting into colour on this tiny mound. It was both enchanting and macabre that this majestic beast should, in death, have given life to the tundra, forming the basis for the nutrition of plants that its own brethren would surely feed on.

As we watched, a huge hairy bumblebee flew away from the buttercups. There are only two bumblebee species here. In one of these, the queen emerges first, lays her eggs and nurtures them into workers who head out to gather pollen and nectar to feed the rest of the brood in this short frantic period of summer splendour. The other species of queen bee emerges shortly after, enters the burrows of the other species, stings the queen to death, then steals her work-force through pheromone entrancement, tricking them into raising her own young.*

---

* This may sound cruel, but interactions between parasites and host are actually the norm, and probably one of the most critical factors in evolution.

We walked up the eastern riverbank for a few hours, crossing endless small streams, trying to make up our minds where we should finally make our assault on the awfulness of the river crossing. There was no right or wrong way of doing it. Eventually the decision was made that we would make a sacrifice of one expendable member of the team. Me. I had brought along a pair of wellies for the occasion, hoping they would at least give me some respite in some of the shallower bits.

I duly waded out into the flow, and to begin with it was all pretty easy, the water barely lapping at the tops of my wellies. In between the channels was mud, some of which was firm enough to walk over, other sections of which were practically quicksand and almost sucked the boots off my feet.

Inevitably, though, I soon waded in over the tops of my boots, which filled up with icy water, instantly freezing my toes. This was just the start. The further I got from the bank and the team, the more I realized how the optical illusion of the Arctic had cursed us yet again. When I thought I was halfway, I looked across and realized I had clearly not even done 10 per cent of it yet, and ahead of me was the main flow, this time broad and deep – a real river.

I briefly considered going back and starting again, getting the crew and some rope to do the crossing properly as we'd been trained. But that would mean retracing my steps, and I just wanted to get it over and done with, so I plunged into the river. Straight away, the water surged over my boots and up to tease the part of the body no man ever wants splashed with freezing cold-water. It got worse, surging up to my waist. I bent double, trying to keep my over-weighted backpack out of the river, as my sleeping bag was right at the bottom of it.

Suddenly the mud on the riverbed caught a hold of my boots, sucking them off my feet and holding me fast. The crew were a mile and half an hour away; I was chest deep in a frozen glacial river and about to die, thanks to a rucksack filled with wildlife textbooks. For the second time in a matter of hours, I seemed destined to perish in a disappointing puff of irony.*

But not today. With a squelch, my boot came free of the mud, and I fought to the other side. It took the team the best part of three hours to get right across the riverbed. When we finally got past all the water, we came to endless flats of mud and sand. Then, weirdly, the sun came out in such blazing fashion that it practically flayed the skin off our faces. With the flat sands and the barren mountainsides, for a short while it felt as if we were striding bare-foot across some alien planet.

A lone plover staggered away from me, cheeping loudly, dragging its wing in a classic distraction display, trying to lead a predator (me) away from its nest and eggs by faking an injury. A long-tailed skua half-heartedly tried to dive bomb me, also doubtless trying to drive me away from its hidden nest. Then suddenly, my footprints

---

* My mum always said that she was convinced, for all my swimming with great whites and catching cobras and crocodiles, I was going to die in the most banal way possible – run over by a milk float, or choking on a cornflake.

came up to a line of far bigger and more impressive tracks. Polar bear, and they were fresh! All of a sudden, the freedom and pleasure of walking barefoot through the river sands came screeching to a halt.

The others were a good kilometre away, and I had no bear deterrents – no gun, no bangers, nothing, not even a flare. My rational mind told me that the prints were a day old, and that the bear had been making a beeline up the river valley and over towards the sea ice to the north, that I would be able to see him a mile away. But my subconscious was suddenly a little boy again – vulnerable, exposed and alone out on the mud flats. It was quite an anxious experience before we could reunite the group and take stock. From there on in, we took more care to walk closer together, and never far from the rifle.

After several hours of comedic squelching, we finished the last few kilometres across the mud flats together. An Arctic fox strode off over the hillocks and hummocks, already wearing its brown summer coat. The changing of winter whites to more neutral summer colours is critical to the survival of animals like Arctic fox and hare. If it is out of kilter with the snowmelt and formation by even a few days, both predator and prey stand out as if they are wearing neon jumpsuits. As the change appears to be synchronized to the lengthening and shortening of days, climate change brings a possibility of a real disconnect. Snow and ice will melt earlier and form later, and this could very literally leave these Arctic mammals exposed.

The walk ended up as an 11-hour march, heading up the

ancient moraine of the glacier we would be spending the next week on. It was steep, and crumbly, perhaps a warning of what would be ahead of us on the climb. Our oversized packs bit into our shoulders, and by the end every step started to become a king-sized drag. Finally we got back to the snowline, and reached the kit dump, which had been dropped off here by the helicopter.

Next morning, we switched to yet another mode of transport, using skis and sleds to progress up the valley and into the high mountains. We were using big plastic sleds called pulks, with a harness and trace system that clips onto the skier, allowing you to drag large loads across the snow. The skier wears skis with a free-heel binding, and 'skins' on the underside of the ski. These have been named *peau de phoque* or sealskins, as they used to be made from the fur of seals. Both the sealskin and its modern synthetic replacements are slick in one direction, allowing movement, but grip the snow if rubbed backwards. Using this system, you can drag even big loads uphill over big distances. It's a terrific method of transporting equipment into hard-to-reach snowy locations.

It is, however, extremely physical work, often said to be the perfect exercise, using more muscles of the body than any other cardiovascular workout. You burn a lot of calories – as many as 6,000 a day – and need to fuel yourself properly, or you can hit a massive wall. Also, the pulks work great until you hit really steep slopes. Then they can become near impossible to move. It was going to need a real team effort to get the half-tonne of gear we had up to our glacier camp. In several places where the terrain got steep, we had to leave our harnesses in order to push our colleague's sleds, working as a unit to try to get everything in place.

Early afternoon, we crested one final sharp slope and found ourselves up on the glacier. Though the terrain continued upwards, it now seemed without features – a sweep of smooth white like a giant undulating freshly laundered bed sheet. On all sides rose hulking rock peaks, hemming us in to an amphitheatre of snow. It was utterly windless, and the sun beat down on us with Saharan ferocity. On the thermometer the temperature would not have been much above zero, but with no wind and no shade, plus the solar glare rebounding off the white below, we were all leaking sweat and down to our T-shirts. Sunblock ran into our eyes, lips cracked and crusted. Taking off sunglasses to clean them even for a few seconds resulted in a piercing squint and the sensation of lemon drops dripping into the eyeballs. Without sunglasses, you could go snowblind here in minutes.

We stopped to survey the peaks and get a sense of objectives.

'I like the look of that one,' said Tamsin, pointing to the most obvious peak – a perfect pyramid at the end of the valley.

'All of these are unclimbed,' Keith affirmed, motioning to the ten or so other mountains around us. The summits were linked by several high ridgelines.

'It blows your mind, doesn't it?' said Tamsin wistfully, 'that even in this day and age you can come somewhere no one has ever been before.' She paused, staring intently up at the mountains. 'It actually makes me quite emotional.'

'Yeah,' agreed Keith, 'and if none of these peaks have ever been climbed, it's a safe bet that no one has ever been up on this glacier before. We're the first people ever to stand here.'

That was true. The local Inuit have no tradition of climbing. Glaciers like these offered no hunting opportunities and considerable dangers. There was absolutely no reason for anyone to have been here before. We all gazed for a few moments at the ludicrous view, letting it sink in. A panorama to match anything in the Pyrenees or Rockies, and our little family was the first to share it.

'I think we should be looking to do a traverse,' said Libby. 'Climb up there or there,' she pointed to several achievable-looking routes that could enable us to get up high, 'and then continue on, bag a few peaks and come down on the other side.'

'We should do all of them,' Aldo said. I blinked at him. He didn't look as if he was joking.

'I mean we have infinite sunlight, right? We should just go, stay up high and bag all of these.' He pointed to each mountain top in turn.

'Yeah, that would be amazing,' Tamsin agreed.

'And I guess we all get to name a peak,' said Aldo. 'I'd like to name one for Anna's dad [Anna is his fiancée] and another for a mate in the Marines we lost.'

I kept my mouth shut, but inside my mind was racing. Was I missing something? These peaks may not have been hugely high, but they were massive objectives. And we'd already seen that the rock and snow were likely to be horrible. To my mind it was going to be an appalling slog just to climb one of these mountains. To be talking about doing them all seemed like insanity. The others were getting carried away.

'Where's our descent line?' Tamsin pondered.

'Probably all the way over here,' Libby mused.

'So we could start on the other side,' Aldo added, 'and do a complete traverse of the valley.'

That was probably ten summits. And though it didn't appear that they'd be technically challenging, I worried we were setting ourselves up for failure.

'Well, you all can do what you want,' I said. 'I'll be coming down that easy ramp there,' I pointed, 'and you all can carry on.' It came out bolshy and confrontational, though I didn't mean it to – I just felt we were in danger of biting off way more than we could chew. There was a moment of silence. The others were clearly digesting what I had blurted, and thinking, 'Steve's scared.'

We got into camp in the early evening, the heat tempered by the sun slipping down behind one of the peaks. We would be camping on the snow covering the glacier for a few days, so it was crucial to make a good flat surface before putting up the tents. This is something I've learned the hard way over the years camping in the mountains. It's tempting when you're exhausted just to whack the tent up, but within an hour or so of lying in your tent the snow compacts into ice. If it's not perfectly flat then every crease and lump freezes into a solid lump beneath you. I started out with a shovel, taking off the powdery upper layer before walking round in ever decreasing circles to create a nice dense platform.

Meanwhile, Tamsin skied round the confines of camp with a long, thin probing stick, plunging it down into the snow to check for crevasses. As the ice from a glacier consolidates and starts to rumble downhill with gravity, it rolls over changes in terrain. Much like a whitewater river forms falls, rapids and whirlpools, the ice cracks, tumbles and gets contoured over the rock beneath. Inevitably, this creates cracks, or crevasses. Some of these on the Greenland ice sheet can be hundreds of metres deep. Up here on this section of glacier they'd probably be no more than 40 or 50 metres, but that is more than enough to swallow you up. This is probably the most dangerous time of year, as the crevasses are still mostly covered over with snow, so you can't see them, but the snow is starting to melt away. Our skis spread our weight. However, now we were in camp we'd be walking around in our boots. It would be easy to penetrate the crust and just disappear down into the darkness.

It was now mid-June, and the eve of the longest day, the summer solstice. This has always been a vital time for local peoples; a time of plenty, when the Arctic cottongrass blooms, when the musk oxen hunting is good, and the geese and eiders arrive in their thousands. The bleakness of winter is a distant memory, and it's a time for celebration and feasting. It is also the time of the midnight sun.

Up here on this high glacier, though, there was enough dip in the trajectory of the sun that there would be a delineated night-time that was different from the day, and though it would be light all night, it would be markedly colder. Therefore the snow would be more consolidated, the rocks should be more firmly held in place by the ice and it should be much safer on the high peaks. For all these reasons, we would climb through the night and aim to be back down at our tents by midday, before the melt made avalanches a certainty.

We left camp at three in the afternoon, having spent the long day packing and unpacking our bags, sharpening our crampons and filling our bodies with as much fluid as possible. It was certain we wouldn't find any running water on the climb, and though I was carrying a stove and pot, we all knew we'd be too busy moving to stop and melt snow. As the five of us skied up towards the base of our first peak, we shaded our eyes from the burning sun and studied the topography, searching for a weakness, for a place where the mountain would more generously give up its treasures and allow us to get up high as quickly as possible.

What we saw filled me with dismay. The snow slopes that had looked so easy from a distance were mostly no-go zones. Almost all of them were run with avalanche debris, and slides were occurring every few minutes, rumbling down with a mixture of snow slurry and rock. In fact, as we got closer, we made the call that we couldn't actually start on our intended side of the valley at all. Its south-facing aspect meant that all of the snow there was just too unstable. So, now we had to swing 180 degrees and try to find a whole new approach. The most obvious place was the ridge we had scoped to be our ultimate descent point.

'It kind of looks like Aonach Eagach,' said Aldo.

Keith agreed. 'Yeah, it looks pretty cruisy.' (Meaning easy.)

Again I shook my head with silent frustration. Aonach Eagach is a steady, classic ridgeline in Scotland's Glen Coe. Even in mid-winter, it's a safe and solid scramble that takes about nine hours. All my mountaineering in the Arctic had been treacherous, and everything we'd seen so far made me think this could be the worst yet.

It was getting on for seven in the evening before we reached the base of the peak and could finally start climbing, still burning hot, sweat and sunblock dribbling into our red-raw eyes. I was exhausted. The last few days of heavy hauling had taken more out of me than I cared to admit – my thighs were aching, my shoulders were chafed and sore. There was nothing I felt less like than a marathon overnight climb. The only way to smash the demons, though, was to take the first lead.

Though the terrain was easy, it was complicated by hideous snow conditions. There was a thin crust on top and below was just mush. You took one step forward, then slowly sank in up to your waist. We didn't rope up on the first snow slope – there was no point. You couldn't have made an anchor in the snow, and if someone had taken a fall, they'd have just dragged their fellow climbers down with them.

After a while, I ditched my ice axe and took on an unusual style of climbing, spreading my weight spider style and punching my fists through the crust to act as an anchor. I must have looked like a baby taking its first crawl ... except upwards. It was laborious, but it worked. I was disheartened, though, to look down and see lightweight Libby just walking up the snow, barely sinking in at all. Being a big heifer like me is rarely an advantage in the outdoors.

It took an hour just to reach the ridgeline and see what we were taking on. There was little satisfaction from the realization that I had been right; we had bitten off more than we could chew. That was tempered with the certainty that we were in for a very long night. The ridgeline was massive and seemed to stretch up forever.

More of a concern, though, was the state of the rock we would be on. Instead of nice solid granite, it looked like the gods had just crumbled up some gravel in their mighty paws and ditched it onto the ridgeline. The rock quality was the worst any of us had ever seen. Nothing, but nothing was solid. Everything you stood on, or tried to use as a handhold, would just fracture and break. It was like climbing on a monster stack of sugar cubes.

Tamsin started to take the lead, and every single step set off a slide of rocks, some of which were the size of microwave ovens. For every single second we were on the peak, we would need to be laser focused, concentrating utterly on the task ahead. If we lost concentration for even a blink, the results would be rockfall, avalanche, perhaps a tumble. Out here, a hundred miles from the nearest human being, this was unthinkable.

Being on the edge for long periods of time is exhausting and debilitating. The effects of adrenalin and the stress hormone cortisol can be to your advantage in the short term. They heighten your senses, re-direct blood from the stomach to your lungs and legs in case you need to run, and put your system on high alert. However, over hours, with an overdose of adrenalin zinging through your veins, you start to feel as if you're being constantly zapped with electric shocks. You're in a state of extreme anxiety. The lack of blood to your stomach, and therefore the lack of a desire to eat, can lead to a plunge in blood sugar levels. It can have weird effects on your digestive system too. In extreme cases, you wee yourself to lighten the load. In less extreme cases, you can get cramps. I tend to get furious acid stomach and stress-induced trumping. This results in high-powered, high-altitude farts of stupendous volume and odour that could strip paint. As warm air rises, the guff gifts were floating up on the breeze to poor Tamsin, who was leading. She gasped and pretended to hurl. But really I reckon she used them as subterfuge to squeeze out a few of her own.

The ridge seemed from below to be split into three main sections, each topped by 'gendarmes', both an old French word for a cavalryman at war and for a pinnacle that stands proud on a ridge-line. Normally these offer the most classic element of a climb like this, with challenging and vertical climbing, and huge exposure (sense of height and drop) to either side. Here, though, they merely marked the places where you could pull the biggest rocks down on top of you.

We were climbing Alpine style, trying to move as fast as

possible together as a unit. As we moved, we'd flip the rope over rocks and use natural protection as much as possible to cut short a fall, should it occur. As Tamsin flipped the rope over one gendarme, and the whole thing just toppled sideways like a wonky statue. I tried to pull myself up on a pinnacle the size of a fridge freezer, and it shifted with a scraping-nails-down-blackboard screech. Every turn sent a cavalcade of rocks tumbling down the snow slopes, triggering little avalanches. There was not a single metre of the climb that offered an ounce of pleasure; it was just pure graft. Our conversations occurred solely to mark the passing of time.

'To anyone that's interested, it's now 2am!' Aldo would announce with a flourish, as we looked around at the greying weedy light. Weirdly, even though the sun was low in the sky, it still seemed as if the snow-covered summits were reflecting back bright white light – almost like those glow-in-the-dark stars you glue on your ceiling as a kid.

Five hours into the climb and each of our litres of water had been long drunk. We sated our parched mouths with handfuls of gravelly snow, crunching the grit between furry teeth. There may not have been any water, but we had brought stacks of chocolate, and devoured it like Augustus Gloop let loose in an Oompa-Loompa's candy cabinet.

'D'you think it's possible to get type 2 diabetes in a month?' Aldo asked.

'I never eat chocolate or sweeties at home,' I replied, 'but here I'm necking it like a Glaswegian at an all-you-can-eat buffet.'

'If it was in Glasgow, we'd deep-fry it.'

'You don't really do that, do you? I thought it was an urban myth!'

'Damn right. When it's cold and wet, you just can't stop eating sugar. And Glasgow is always cold and wet.' Aldo was matter-of-fact.

'Speaking of which,' I mused, 'I have just done a wee that looked like Irn-Bru.'

'Word, me brethren,' said Aldo, doing an unconvincing London accent. 'I'm coming down from this mountain looking like one of our freeze-dried meals. You'll have to top me up with boiling water to rehydrate me.'

Eventually, as the warmth started to return, the snow to glisten with diamond crystals and the sky to turn from pavement grey to baby-blue, we reached the last crunchy ice slope before the summit. A snow crest like a breaking wave led towards the actual top. This is known as a cornice, and can be quite dangerous – liable to break away or crack apart. Particularly as in this case it had never felt a footfall before. By this stage we simply couldn't have cared less.

'D'you know what?' Keith said. 'I think this view is better than from the summit of Everest.'

It was a fitting sentiment for the moment. The mountain tops around us seemed to be close enough to touch, stretching off to the horizon 50 or 60 miles in every direction. We could see as far as the boundaries of a decent-sized country, yet we knew for a fact there was not another human being in that landscape.

To the south, we could see all the way to the fjord, the giant bergs in Iceberg Alley from here looking like tiny distant sugar cubes. To the east, we could see past Schubert Dal, and down the entire route we'd spent the last week battling through. North and west were just mountains. No one stopped to count how many peaks we could see, but there must have been several hundred, and a decent amount of those remained unclimbed. It looked like a lot of adventure left to be done.

Tamsin was first to summit, walking on the crest of the wave. She tested the ice with her axe; it seemed secure. Stepping exactly in her compressed footprints, we made our way across to join her, where we all hugged, slapped backs, shook hands and savoured the occasion.

When you make a first ascent in this day and age, it's not done to talk about conquests, and conquering a peak. The mountains tolerate you at best. On another day, in different conditions, the summit you've just won could have kicked your behind, or even killed you. The summit was an emotional place for me. The language might have changed, but it is still tradition that you get to name your peak. I'd done a little bit of work, asked some questions of the elders in Ittoqqortoormiit, and gained their permission to name the peak should we summit. The rest of the team had gladly agreed that I could have the honour, and for me there was only one option: Angajullilaq. The local word for 'first-born son'.

The descent was just as sketchy and scary as the climb had been. All the way my mind was full of thoughts of home, of Helen struggling with builders and bills while heavily pregnant, of the fact we had not spoken in weeks. My mental grouch made the descent a trudge, though it thankfully finished with an exhilarating ski back down to camp. It was noon by the time we were back, after 21 hours of non-stop effort. None of us had drunk more than a litre of fluid plus a few handfuls of snow, or eaten anything other than cereal bars. We were utterly exhausted, and collapsed gratefully into our tents, slipping into our sleeping bags as if they were four-posters in the world's finest hotel. I slept for 16 hours.

We reached the foot of the snowline on a miserable evening, piddly grey rain soaking us as we put our tents up, permeating every item of clothing and every ounce of kit. The forecast showed this area of low pressure was set to stay for the next few days. The chopper would never fly in this, meaning we were condemned to fester here, low on food, psychologically finished and ready for home.

Spirits were low, but thankfully we had set that evening aside to celebrate Rosie and Keith's joint birthdays. Keith had kept hold of a bottle of Highland Park whisky and I'd brought a bottle of ruby port out for the occasion. I also had two cakes from my local farm shop. The only question was where to get candles.

I ferreted through the rubbish and found some red wax from some Babybel cheeses and the remnants of a ChapStick. I melted both of these onto some threads from a piece of old rope that I had soaked in some of our cooking fuel. Candles! We stood around in the glum clouds, playing pass the parcel, doing silly forfeits, telling jokes, singing songs. It struck me then as it often does that the moments I will treasure more than any from these expeditions will not be the summits or the sunrises, it won't be achievements or heartaches, it will be the friendships. It has been the great privilege of my life to work alongside some incredibly talented, kind-hearted, dedicated, wonderful people. There are times when we can chat about how much we miss home with a candour that would never work back in the real world. When we can voice opinions, talk about stuff that matters and lots that doesn't. When a simple bar of age-whitened chocolate that has melted and reconstituted one too many times can be the greatest luxury you could ever wish for. When a simple dram of Scotch can taste like nectar of the gods.

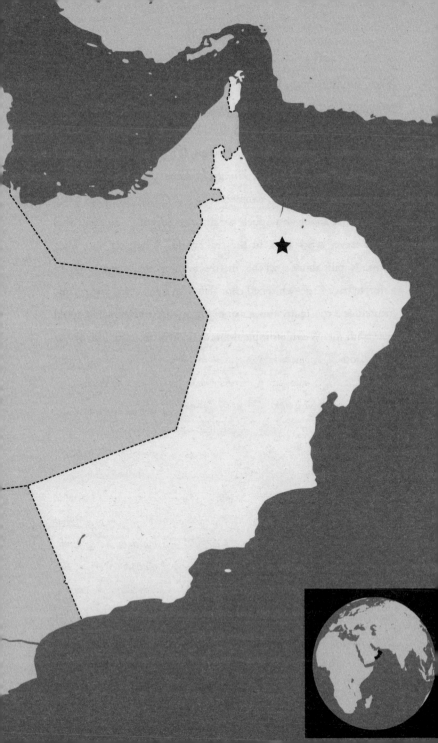

5. DESERT CANYON

OMAN

Fierce white light sizzles on the skin like acid, burns bubble and blister on my neck and the backs of my hands. A salt crust ebbs over our clothes in swirls like monochrome tie-dye, chemical sun blocks leak into the eyes, leaving us with the red-raw gaze of desperate men. It is a long way down. How else to describe such a drop? The only equivalent in the man-made world would be to stand on the top of the Empire State Building's radio mast and step off into the abyss. And this was what we were about to do. Great height is a universal fear that every mammal is born with. Even a goat kid, puppy or kitten will avoid a great drop from instinct alone. This, then, is a battle with primal panic. I have a long hard word with myself, telling myself that the little-finger-width of rope will hold, that I will not drop to my death, that science stands between me and the hot rocks half a kilometre below. But the sum of science is paltry in the face of fear. My body is telling me that the step means certain, horrible death. My mind tells me to sort myself out and let go. I straighten my legs, lean back and propel myself into the unknown.

It was a mere six weeks between the Arctic and the desert, but it might as well have been a lifetime for me. This was the only time I'd set aside in the whole year for Helen and me to have our baby. What to write next? It's so difficult. Why is it I find it easy to talk about crocodiles and first ascents, but so hard to write about the important things that people do all the time?

I rushed back from the Arctic like I was fleeing a burning building, every second that I wasn't travelling towards home feeling like it was criminally wasted. As it turned out, that rush was merely to find myself in NCT classes, learning about baby-weaning, nappy changing and nipple rash. Shortly after my return, though, the doctors decided that with the troubles Helen had been through in pregnancy, we needed her to be induced.

As is typical of my over-achieving wife, the previous day we went for a long kayak down the river and had a barbecue by the house-boat, where we sat late into the evening, talking genially about how this was the last time we would ever do this as a couple alone together. The next morning we rose early, with bags packed as if heading for a spa weekend, and drove to the hospital.

It seems my relationship with Helen has been all about destroying my inner control freak. Watching her compete at the Olympics was a horror. Seeing something play out that means *everything*, but able to do nothing about it . . . it was the most frustrating experience of my life. And now here was something even worse. For 24 hours of labour I had to watch the person who means more to me than anything go through agony, and all I could do was rub her back. After eschewing pain relief for all that time, finally at the last second things had to get surgical, and she had a gargantuan needle injected into her spine in between contractions. I had to sit there like a lemon holding her hand and praying to any deity that might hear me to take care of her. But then the following morning, to have a bloody seal pup of a baby placed on Helen's chest in front of me . . .

Many male friends had talked to me before about the bewildering lack of empathy they felt when their baby was born, the disappointing lack of connection, the empty sensation of knowing you should be ecstatic, but feeling nothing. That didn't happen to me; I had the thunderbolt. The second my son arrived in front of me, everything in my life changed. All I wanted to do was snatch him from the midwives and back out of the door, clutching him to my chest. Driving home, I could have physically assaulted anyone driving too close. Walking down the street with him, I had to remind myself to unclench my fists and relax my jaw. I saw danger everywhere, saw terrible potential in every passer-by, felt possessive fear only for him.

We'd decided long ago to name our newborn son Logan, after my best friend, and also our happy place Logan Rock down in Cornwall. Everything about those first few months was a surprise to me. I was totally blindsided. It's alarming how completely the animal brain commands us as modern humans. Before having Logan, had anyone talked of how we are governed by ancient impulses they would have seemed like some kind of tin-foil, hat-wearing mentalist. Yet I now realize we are only starting to scratch the surface of how servile we are to a mainlined hit of hormones.

Suddenly, Helen – who as a professional athlete needed nine hours' sleep a night – was thriving on just two. Her entire focus had become to provide, mine to protect. The primality of it was fascinating to behold. Physically too, the changes were intense. We both smelt different. You could scent the heavy musky aroma of parenthood on us both, and Logan could too, long before he was really aware of anything tangible around him. Sights, sounds and smells all inspired a torrent of hormonal drugs, coursing through our system, dominating our actions, moods and thoughts.

As someone who studies animals for a living, it should be no surprise to me that we have not evolved beyond the confines of our biology just yet. As Desmond Morris, author of *The Naked Ape*, famously wrote: 'We are risen apes, not fallen angels.' Our sense of smell is enough that a man can scent a receptive woman in a gathering of hundreds. Our hearing is so acute that blind children can learn to echolocate like bats. The fine hairs that coat us could be used to sense vibration, wind movement and bodies moving about us. We almost certainly perceive the world's magnetic field, and could use it like all migrating animals do . . . if only we didn't obscure it with the trappings of the modern world. We are just animals, and potentially as capable as our mammalian cousins. We just don't realize it. Well, until moments when the drugs of our hormones are truly pumping around our bodies.

The hormone you hear about most in pregnancy and childbirth is oxytocin. Many would say it dominates everything in the process and is the chemical that leads us by the nose to being good parents. And it turned me into my baby's bodyguard.

Journalist and author Jon Ronson wrote of his irrational fear on being away from his son. He always feared something appalling had happened to him, envisioning all the ways his son could be hurt or worse, so much so that it became a serious condition dominating his life. I had the same thing. Every night I was away from Logan and Helen, I was plagued by nightmares of terrible things happening to them. Always gory, always dramatic, and I was always separated from them by an invisible wall, seeing all, unable to do anything about it. And now I was scheduled to spend four of the first five months of my son's life in a jungle, up a mountain and in a cave, not even able to make a phone call to them. I was in for a rough ride.

The gulf below us was pure blackness, accentuated by the searing white heat of the rock above. Kids with big black eyes in ankle-length white robes played around the edge of the hole as if balancing on a playground wall, rather than teetering over a drop of hundreds of metres. Uninterested camels chewed sandy cud under the petty shade of nearby acacias. Shaggy goats with weird horizontal pupils sneaked in while we weren't looking and tried to make off with our sandwiches. Meanwhile, all of us were hypnotized, staring down into the big black hole that we were preparing to abseil into.

Aldo had been rigging a rickety tripod for the last four hours, tying it off to the wheels of our Land Rover. The contraption was designed to push our ropes out from the sharp limestone, which would otherwise saw through them under tension. He'd been working in relentless sun and looked like he'd showered in his clothes. Every time he leaned forward a shower of sweat leaked from under the rim of his helmet and dribbled through his beard.

As I backed over the edge, the familiar plummet in my stomach dropped, as if my guts were making the journey down before me. You'd think after all these years I'd be immune to this sick feeling, but it never goes. I guess it's six million years of evolution telling my body that stepping off into a gaping drop is a dumb thing to do. I stared into the abyss; it stared back up into me. It's that first step that's the worst, where you leave the security of your feet on solid rock and just let them swing. It's a horrible sensation. But then I was on the ropes and I was falling.

I took my eyes off my descending device and risked a look around. The hole was probably as big as a tennis court and the desert floor had just fallen away, like those sinkholes that sometimes open up in suburban streets and swallow a family sedan. Below the ground, though, was a hole that would need more than just a few blokes in high-viz to fix.

The ceiling curved away. As my eyes became used to the darkness, the curve went on and on. It was how it felt to be inside a mountain, like someone buried Wembley Stadium and just forgot about it. This was not the only hole in the ceiling of the cave; there were two others, and as the midday sun poured in through the gaps, it sent down perfect shafts of sunbeam. Majlis is a sinkhole in the making. Eventually, these skylights will crumble, the whole roof will cave in, and this arena will be exposed to the desert sun. That could happen in one catastrophic event, or it could take millennia.

I was descending through one of these spotlights, though that was only visible from way below me, deep in the cave, when I shouted 'Day-O!' into the blackness. (Strange that Harry Belafonte should come to me at a time like this.) 'Day-O!' the cave shouted back, again and again. It was the greatest echo I'd ever heard, and took at least ten seconds to peter out.

Majlis is one of the largest natural chambers in the world, certainly the largest in the Middle East. Well, the largest we know of anyway. The nature of this landscape is such that there are almost certainly untold wonders like this littered through these mountains that we just haven't found yet. The cave was discovered in the 1980s by surveyors who were flying aerial recces over these mountains, searching for signs of subterranean water. Because of the punishing heat of the desert, surface water doesn't last long here, and all the water is contained in aquifers deep below the surface. As such, places like these are the lifeblood of any civilization in the Middle East. Arabian peoples once believed that caves were the homes of jinn, spirits or genies, so when this place was first discovered, they named it Majlis al Jinn, 'the meeting place of the spirits'. It is surely one of the world's natural wonders. Stretching 300 metres in length, it's big enough to accommodate 12 jumbo jets, wing tip to wing tip.

Originally, all of this rock around me was sea-bed, corals and the skeletons and shells of long-passed marine organisms. Then tectonic squeeze hoisted this land to over 1,000 metres above sea level. The evidence is the countless marine fossils found all through this rock, and even resting on the surface of the desert as if they've just been dropped there.

My descent into the gloom was going fantastically well until we decided to fly our drone over the top of the hole to look down into the blackness. Almost immediately, two beaten-up flat-bed trucks rumbled out of the nearby shacks at top speed and screeched to a halt by the edge of the hole. Three men emerged, shouting angrily at my crew and gesticulating towards the drone. My Arabic only extends to the peremptory greetings, but it was totally clear what they were saying: we had flown the drone up where it could look down into their houses, into their yards. Where it could film their women and children, and worse, film down on their heads. In Islamic lore, the head is the most sacred part of the body. In many Islamic countries, it is the worst possible offence to pat someone on the head, or conversely, to point to anything with your foot, as this is the lowliest part of the human form.

It took us a good couple of hours of placating before they forgave us, at which point they brought out plates of dates, fruit and coffee to share. We went from pariah to honoured guests in seconds! Popping the first date in my mouth is the taste experience of a lifetime. Dates to me are the dried ones we get at Christmas, wrinkled like post-bath fingers, I'd never had them fresh off the palm tree. These looked like fat golden acorns, but tasted like a mushy explosion of butterscotch and caramel. The coffee was sweet and scented with cardamom. This kind of interaction is normal here; it's socially sanctioned for emotions to run high, but the hospitality is humbling.

It's half the temperature down in the cave as it is above. While the desert was over 40 degrees centigrade, down there it was only 20. The further I went, the more my eyes became accustomed to the gloom and the scale of Majlis became ever clearer. Finally, my feet touched ground, 180 metres below the surface.

Off to one corner of the amphitheatre was a dry lake, the ground still sodden from the last rains. After a big deluge, this area apparently fills with water that quickly percolates down through the surface, running off into cave systems, which eventually carry the subterranean watercourse all the way to the sea. After just a few hours, I needed to drag myself up that rope back into the light and the heat, because the aim of the expedition was to follow this passage of ancient water into places no human being had ever been before.

Oman is a nation that has been calling for many years. My first professional expeditions were in the desert canyons of the Middle East back in the 1990s, and the majesty and ferocity of these landscapes will always hold a fascination for me. Oman is part of the Arabian peninsula, but not as restrictive or war-torn as many of its neighbouring countries. It is a nation of tantalizing beauty, windswept exoticism and, most enticingly, wild canyons yet to be explored.

My contact there was Khaled Abdul Malak, a surgeon from Lebanon, who now works as a dentist in Muscat and has lived there for a quarter of a century. Khaled had explored 16 new canyons in the mountains of Oman, and had a handful of new ones he'd been saving for a (hopefully not) rainy day. He agreed to share with me that privileged information and show me one of these 'undropped' canyons, somewhere in the mountain range of Jebel Akhdar.

The tricky thing was getting Khaled to open up about what this expedition would actually entail. It wasn't new to me for my contacts to be cagey about their information. Many explorers cling to their potential 'firsts' jealously, for fear someone else might get there before them. The annals of explorer history are littered with great races to poles, planets and pinnacles, and there are endless tales of prizes pilfered and glory gazumped. However, in this case, it was only on the night before we went into the canyon that Khaled shared with us even the mountain we would be heading to. He might as well have forced us to wear blindfolds so we wouldn't be able to unveil his secrets to anyone.

My other friend in the country was Justin Halls, who was with Aldo and me for our first descent of the Baliem River. British by origin, he had married an Omani girl, and found a better life in the dunes and dry river valleys of Arabia.

Driving across Oman to our first destination was an adventure in itself. The country has reaped the benefits of its substantial oil reserves and appears to have spread that wealth around more fairly than many other oil-producing nations. Muscat is a city of huge glittering mosques, with an opera house lush with Italian marble and vast shopping centres, all framed by craggy mountains. It doesn't have the overpowering opulence and skyscrapers of Dubai, and retains much of its historic flavour in the souks and alleyways of its old town, thick with the heady scent of burning frankincense (as is practically everywhere in Oman).

The roads are mostly six- or eight-lane slick highways, and a joy to drive – unless it's early afternoon during Ramadan when everyone is busting to get home and suffering from critical blood sugar levels. It would be one of the world's great transport networks, were it not for a Lexus- and Bentley-driving mega-rich class, who drive to a different highway code to the rest of the country. You'll be gliding along nicely when suddenly a ¼-million-dollar car will undertake, cut up in front of you, ram on the brakes a metre from your bumper then accelerate away, doing 120. The first time this happened, Aldo roared in anger, throwing his hands in the air.

'Don't do that!' Justin squealed. 'You'll get arrested!'

'Do what?' an incredulous Aldo responded.

'I'm serious. Making angry hand signals is an offence here. I got locked up for two days for flipping someone the finger when they cut me up.'

'What about if some douche in a flash car nearly kills you?' Aldo continued.

'Suck it up,' Justin said. 'Anything that could be thought of as road rage can get you banged up.'

We resolved to be more sanguine. An hour later we pulled off the shiny tarmac and onto off-road gravel. Soon we were driving up steep mountainsides, with cud-chewing camels eyeing us with a typical lack of interest.

Justin's local partner was Abdul Hamid. Tall and slender, he has big black eyes and an easy manner. Except when he's talking to Justin. Then all bets are off. Even keeping them in separate vehicles didn't solve the problem. As we drove through the desert, the two of them bickered on the radios like an old married couple.

'Hamid, will you slow down! You're going to hit a rock and take out the bottom of the vehicle.'

'I'm doing, like, ten miles an hour! You should stop driving like my great grandfather.'

'I'm staying back to avoid the massive cloud of dust you're kicking up.'

'Well, if you can see my dust, just follow it and you won't get lost.'

'I'm not worried about getting lost. I know these roads like the back of my hand.'

'You get lost in the mall, even when you have GPS.'

And on and on, all the way to Jebel Shams.

Before we began the exploratory leg of the trip, Khaled was

taking us to Jebel Shams, the highest peak in Oman, at 3,004 metres above sea level. It is itself the biggest peak in the Jebel Akhdar range, an intimidating, parched massif. We spent the first night in a hotel towards the summit of the mountain – there's a military base on top of the peak, which prevents you actually getting to the very top. The summit, though, was not our objective; that we would see soon after sunrise the following day.

In a howling hot gale we walked to a roadside viewing gallery and got our first view of our canyon from behind safety railings. It is another of the world's natural wonders, often known as the Grand Canyon of Arabia. Dropping a vertical mile from its inception to its conclusion, Wadi Naqab is a commanding sight. The rock comes in shades of straw yellow and satsuma orange, with a bit of black and brown thrown in for good measure.

Oman is famous amongst geologists for being one of the world's great repositories of ophiolite – the kind of rock normally found 70 miles beneath the surface where the earth's crust and mantle meet. These rocks sparkle green and purple, an impossible sheen that makes the mountains look like the set of an alien planet from the original *Star Trek*. Here, at the lookout, blown and blustered about by a planet-sized hairdryer of a wind, we gazed down on towering mountains, falling into cracks, crevices and dry riverbeds, many of which were too narrow to see into, even with binoculars. It looks like a big adventure. Or a whole heap of trouble, depending on your mood.

Though there are countless gorges running down Jebel Shams,

it was the central slice we were interested in. It is the Everest of canyons, yet has only been descended one time – by Khaled, with one friend. Our first goal was to be only the second team ever to travel down its length. We hoped a world first would come later with another completely virgin descent, but in the meantime this was the biggest, the most epic, the most dramatic. Even just to get started, we would need to construct the biggest abseil any of us had ever done, with complicated systems and purpose-built equipment. And all of it would need to be done in blinding sunlight and 45-degree heat.

The geology here in Jebel Shams is also uplifted marine limestone, which has been sculpted over millions of years through simple erosion by water. Presumably in the past this was a much less arid place, and constant rivers would have flown down from the mountain tops. The rivers plummeted over many waterfalls and ran through gullies on their way out to the coastal plains and the sea. It's hard to believe that looking at the view now. You'd scarcely believe a drop of rain would ever fall here. When it does, the dry ground and bullet-hard bedding planes mean these dry riverbeds and waterfalls flash flood with water. Biblical amounts of water.

One of the porters showed us a video on his phone of the last time this had happened, just the month before. After four hours of rain, every single dry fall in front of us was transformed into a tumult that would have had Noah scrabbling for building materials. It was enough to make us heed all Khaled's warnings about not sleeping in the wadis, and to keep us constantly checking the weather.

To get down to the start of the descent, we followed the well-trammelled balcony route that hugs the hillsides. As we were carrying over a kilometre of rope, and gallons of water, we hired porters and two donkeys at the head of the trail. Normally, we refuse to use beasts of burden on our expeditions. In many developing nations, mules and donkeys are beaten and starving, overloaded, covered in ticks and fleas. Their lives are miserable and using them to carry our equipment is not something I'd usually countenance. These animals, though, were in beautiful condition, with glossy coats and mascara eyelashes. Aldo and I were also carrying heavier loads on our own backs than the donkeys, and the wisdom of bringing them was further questioned when I found myself walking in between the two of them. If I slowed to tighten a strap or look at the view, the donkey behind me would make to bite my backside, and every few minutes the donkey in front would let rip with a monstrous trump, somehow always timing to when I was breathing in. 'I had a patient like that when I was a doctor,' Khaled commented. 'An old lady, she came in complaining of "humid farts". I'm afraid I laughed so much, she left the surgery before I could suggest a cure.'

The balcony walk must rank as one of the most scenic trails in Arabia, with rounded rock ceilings often looming overhead. Every corner reveals another vista of splendour. After a few kilometres, we came to a vantage point that showed the full canyon we would spend the next few days descending. It began with a monumental archway – a shelf with an undercut below it that put me in mind of a Victorian fireplace. We all walked out to the edge and studied the route through binoculars. I have spent days staring up at rock routes when mountaineering; this was the first time I'd ever done it looking down.

'Do you see any water?' Aldo asked. Other than flash floods, this was our primary concern on the expedition. We were in Oman in the peak of summer, mid-40s centigrade was expected, and it could get hotter than that. There would be less standing water in the desert canyons than at any other time of year, and doing the intense activity we were planning, we would need up to ten litres of water each a day. You can't carry that much, so unless there was going to be natural water in the pools below the dry waterfalls, then we would be in big trouble.

'Yes!' I replied. 'There's definitely a pool just below the first big drop. And another just below that. After that, the canyon drops into a tight gorge. I can't see inside it.'

'There will be water there, I am sure of it,' Khaled stated. He was sure of everything. Almost all of his sentences ended with affirmations of complete and utter certitude. I wished I had half of his confidence.

'It looks big,' Aldo said with typical understatement. 'I'd guess like about the height of the tower in Canary Wharf.'

'Four times higher,' said Khaled. 'Actually, she is taller than the Tour Eiffel.' For the first time his French Lebanese heritage was evident in his speech. 'Three hundred and eighty metres from the lip to the bottom, the same as the Empire State Building.'

'What's the biggest single-drop ab you've done before?' Aldo asked me, using the slang term for an abseil.

'Kaieteur Falls in Guyana. That was about 200, I think.'

'So, half this? Yup, I'm about the same, off a building when I was rigging. What about when we did Angel Falls?'

'Well, that was higher, but our biggest abseil on that was no more than about 80.'

'Amateur hour next to this,' Aldo said. 'And my gear got so hot, it nearly caught fire on that.'

The end of the balcony walk was a deserted village at the top of 'Fireplace Falls', which we would thereafter refer to as the Grand Arch, the 380, or simply 'the Big Drop'. The ruins consisted of dry terraces cut into the steep hillsides and dry stone walls protected by the natural rock ceilings. The watercourse would have provided enough constant supply for villagers to keep the terraces irrigated using the ancient 'falaj' system, where access to water is meted out like credits to each farmer. Even today, 50 years after the village was abandoned, it was a splash of verdant green in the parched landscape. Trees were hung with ripe pomegranates, which we plucked and ate, meagre pools flitted with bugs and rang with the calls of Arabian toads at night.

'This had been occupied for over a thousand years, and until the 1970s,' said Khaled. 'Then the government paid the people to move out and into modern villages. Now that water is such a problem, I bet they wish they'd stayed.'

'You wouldn't want to bring your children up said Aldo,

shaking his head. 'Can you imagine how terrified you'd be, knowing they were playing with that to one side the whole time?' He gestured to the sickening, dizzying drop.

'You get used to it,' Justin said. 'There's a climb just up there, which they used as a shortcut, up and out of the village. By modern British climbing standards it's rated E2 [that's about my climbing limit], and they used to do it barefoot and carrying goods for trading. These people made mountain goats look clumsy.'

'Plus, I guess it's kind of got everything,' Aldo suggested. 'You've got water, shade from the sun, and no enemies could ever sneak up on you here. It's a natural fortress.'

'What's this?' I was looking bewildered at a wooden door to one of the stone buildings. It was clearly hundreds of years old, carved with beautiful designs, an antique work of art – except it was spray-painted with bright green letters in Arabic. The walls around it had also been graffitied, with some big arrows pointing into the door.

'Oh, that?' said Justin. 'That is a sign that says "house". And that one there says, "this way".'

'Seriously?' I was gobsmacked. 'Who would do something like that?'

'A local guy,' Justin said. 'He wanted to be a guide and figured this would be the way to sell himself.'

'Well, I hope the local antiquities people found him and got him arrested?'

'They didn't have to work too hard. He put his name and mobile phone up on the rock wall as well.'

That night we made our camp on the top of the terraces. It was the best campsite you could ever ask for – flat, as if it had been rolled out (which it had, a thousand years ago), and with a view down the canyon we would descend the following morning. As the sun started to lower, a warm wind took the edge off the scalding heat and the air around us filled with dark, spiralling shapes; Egyptian vultures, one of the most acrobatic of all vultures. With binoculars, you could see the ruffling of white and black feathers on the leading and trailing edge of their wings, agitated by the movement of rising warm air currents or thermals. Vultures are so adept at seeking out these pillars of warm air and using them to their advantage that they are believed to use their phenomenal eyesight to look for rising specks of dust, bugs and the ruffling of leaves in trees far below them – all signs that an energy-saving thermal is forming.

That evening was one of those precious moments that make the desert the finest place in which to camp out. You've been searing hot all day long, but then that heat mellows – often to such perfection that you can sleep out with nothing but a sheet between you and the firmament. At this altitude, it chilled enough for a sleeping bag. The falling sun stains the world gold, orange, ochre and then purple – every peak off into the distance stacked lighter and lighter in shade.

We gathered firewood from under the persimmon and acacia trees. It was so dry that one single flick of the flint on steel and it sprang into flames. Trees grow slow here; some can be thousands of years old. Once alight, desert wood is so dense that a stick as thick as my thumb will burn for an hour. I'd forgotten quite how intense a fire you get from these timbers. We couldn't sit anywhere near it. When the embers finally settled, we gathered as close as we could bear, cracked open a bottle of port and, in the flickering light, made our plans for the following morning's descent, which would in effect just be training for the undropped canyon that was to follow some days later.

We'd talked so much before about the technical challenges of rigging a drop this big that I don't think any of us had really taken in how serious the canyon below might be once we actually got into it. In my head, we were pretty much doing the drop then yomping out the same day. However, it was clear from looking through the binoculars that there would be a score of other dry falls, some of which would be close to 100 metres in height. It would be time-consuming, and worst of all, once we had started descending the ropes, there would be no retreat. There were unclimbable cliffs on every side, and we would be locked into the canyon with no chance of rescue. It was formidably committing. Our cameraman Graham, soundman Parker, and Rosie had all completed industrial rope work training in order to join me on the descent. They're all tough and experienced, but even lowering into Majlis al Jinn cave had been the biggest abseil they'd ever done. They were not concealing their nerves.

'So what happens if the wind gets up and you're already descending on the ropes?' asked Rosie.

Aldo whistled. 'That would be all bad,' he admitted. 'Even on the cave we had massive problems with the lines tangling; you do that over the big drop and no one's coming to save you.'*

'I think it'll be worse in a big wind,' I cautioned. 'If it was blowing as much as it can do here, you'd be tossed about like a rag doll. It would take hours to get down the lines, and you could have a really bad day.'

'And then if we take all of us down,' Rosie reasoned, 'it will take all day just to do the big drop, and then how long to get out?'

'Anyone's guess,' I said. 'Three or four days?'

'You know what I'm like,' Graham said, ever the chipper Bristolian. 'I'll follow you anywhere. D'you remember that helicopter in Australia?† But this just looks nuts. I mean, I'll do it, but it's nuts.'

It was a sombre team that rolled out their sleep mats, all preoccupied with whether they should stay or go. For me, the greatest concern was for Rosie, Gray and Parker. They're all great friends,

---

* Our exit from Majlis cave had been a nightmare. The plan had been to winch people and kit up the 180 metres of rope to the surface once we were done, but the ropes had jammed, leaving me suspended halfway up with my own bodyweight in filming equipment attached. I was on the ropes for an hour and a half, and was in a lot of pain by the time I finally climbed out. Justin and Graham came after, and weren't topside till late at night, dehydrated and exhausted.

* We were dangled underneath a speeding helicopter, dropped onto crocodile nests while the female crocodiles snapped at our heels. Every bit as mental as it sounds.

and this was supposed to be a special experience for them. The last thing I wanted was for them to be frightened or worse, hurt.

The following morning I was roused by a goat nibbling in my food bag by my ear. He was a magnificent billy, with thick hair like an Afghan hound, and clearly used to stealing scraps. I had fed him an entire bag of banana chips before Rosie woke and reminded me that those were my expedition rations and I might need them.

'Clearly knows who the soft touch is,' Aldo yawned. 'I'd have put him on a spit and had breakfast kebabs.'

After dehydrated porridge and desiccated blueberries, we wandered down through the terraces to the top of the big drop. For the second time in just a few days, my guts dropped out of my belly. The only way you could bear to go near the edge was lying flat on your stomach, crawling over the shiny limestone to look over the drop. As we did just that, Khaled walked right down to it, and wandered about without a care.

'Khaled, please don't do that,' Aldo pleaded. 'You're making me feel sick.'

Figuring out how to rig the abseil was a substantial logistical challenge. Most of my experience is rigging simple abseils in the mountains using boulders and trees as 'natural protection' to fix myself to. Justin and Aldo, though, are industrial riggers, qualified to rig ropes off skyscrapers, cooling towers and wind turbines. I was outranked, and relegated to menial tasks like coiling and uncoiling the 400-metre ropes. In between these jobs, I retreated to the shade as the burning sun coruscated the rock basin at the top of the drop. Justin and Aldo slaved on, barely stepping into the shade, often hanging out over the drop for hours on end, trying to get the rope system perfect. It was essential that the ropes didn't rub over sharp limestone. When loaded, even the tough cord we were using could rip apart with a couple of sawing motions.

It was a long day for them. The arch was already starting to come into shade by the time the ropes were ready, and both Aldo and Justin looked as if they'd been parboiled. At one point, Justin dislodged a football-sized boulder, and it dropped all the way down to the valley below. It took 12 seconds for the thunderous sound to come back up to them.

With a drop that big, we needed specially designed abseiling devices; essentially a hefty block of metal with a sequence of rungs that you laced the rope through to slow your descent. A normal descending device would get so hot from the friction that if you stopped moving, it would burn through the rope. These blocky descenders were just a big heat sink, spreading out that friction heat and making sure we didn't sizzle our own safety lines. The kit was cumbersome and clunky, and I had a huge bag beneath me filled with ropes, water, and camping and filming kit. All of a sudden this didn't seem like such a great idea.

All around me, vultures circled, riding the hot winds as they blazed about the arch. 'What do they know that we don't?' Rosie asked.

I stepped back, lowered myself down the short face and came to the lip. Below my feet was just empty space, and two ropes disappearing off into the void. As so often in moments of fear, it helps to focus on something else. So I set my mind on getting my systems together. In front of me, the architecture of the arch was revealed – an undercut ceiling lined with stalactites, and green with moss and ferns. I felt the wind catch at me and almost start to lift me back up again.

Once I'd got over the horror of the edge, the cliff face whizzed past me, and the slight spin on the ropes gave me alternating views down the valley and into the gulping bowl beneath the arch. About halfway down, the weight of the ropes below me lessened, and I started to pick up speed. My stopper knot was slipping on the rope, and not slowing me down. The only way I could stop was using my leather climbing gloves, and wrapping the rope around my legs and boots. I added a couple of carabiners to the system, and it slowed me somewhat, but I was still shooting towards the ground far faster than was sensible.

It was less than 20 minutes until my feet touched ground. My work gloves were worn through at the palms, but I was down, and I was alive. Snatching myself off my ropes, I scrambled onto safe ground and struggled to take in my new environment. I was a gnat in the grate of the giant fireplace. The two lines whipped in the wind, all the way up to the ceiling, 400 metres above. All around me was a boulder field, the historical remnants of the feature I'd just abseiled down. Throughout the mess of rocks there were embla-zoned splashes of colour – oleander bushes in pink and white flower. The sun was low in the sky, and leaves swirled around me, carried on the thermals. Except . . . no! They weren't leaves, they were bats! Bats that weren't flying as I'd ever seen before; they were gliding on the updraft like swallows, barely flapping their wings, the full moon highlighted through their outstretched wing membranes.

I picked my way down through the boulders onto safer ground, feeling like the last man on earth in a post-apocalyptic thriller. No more than a mile away as the vulture flies from where I was standing, I knew tourists would be looking over the railings down into the canyon, with no clue about me and my little world.

Out in front of me, the valley narrowed and constricted into the beginnings of the gorge. The boulder field appeared to turn into a smooth flowstone flume, like a giant water slide emptied out for cleaning. I could see the first of the green pools, and nice flat rock ledges where we might find our ideal camp spot. It was beautiful. And I was only the third person in history to see it. The silence and solitude were overwhelming, almost intimidating, but I didn't want to share them. I switched off my radio just for a second, to have that place, that view, all to myself.

An echo rumbled down from above. I switched the radio back on again.

'I'm on my way, Stevo,' Aldo said. 'Best keep out the way of the bottom in case I kick something down.'

After agreeing with Aldo that the film crew would stay above – it was too late in the day to get everyone down – I arranged the ropes at the bottom to give him the best chance of a smooth descent. Then I clambered down the rocks towards the flatter ground.

My bag was half my bodyweight at 45 kilograms, and on the boulder field slopped about on my shoulders alarmingly. I was going to have to do some serious repacking in the morning. By the time I got to safe ground, I looked up, expecting to see Aldo zipping down towards me, but it didn't look as if he had even begun to descend yet. The sun was setting, and light in the canyon was fading. We would have to get moving if the four of us were going to get down before nightfall.

'Aldo, Aldo, it's Steve . . .' I spoke into the radio.

It was two or three minutes before he got back on, and he was puffing and panting: 'Dude, how did you get enough traction out of the descender? Mine's just not stopping!'

He sounded different to how I'd heard him before – edgy even panicked. Aldo would normally amputate one of his own limbs before admitting he had a problem.

'Are you alright, mate?'

'No, not really. My bag's too big, I can't slow myself, and I can't lift it off my system. I'm really struggling.'

So he had two problems. First, he'd been going too slow and so had removed some of the friction, and now he was going too fast and couldn't stop himself. This was not good.

'Aldo, mate, you need to generate more friction,' I said. 'Can you put another biner high on your system and run it up through that?'

'Tried that,' he gasped. 'I'm just fricking stuck.'

I looked up the lines, thrown about like the rigging of a schooner in a squall. Somewhere up there my pal was battling against elemental forces – against friction, the dwindling energy stores in his muscles, heat building throughout his equipment and body, and gravity threatening to throw him down the line at suicidal speeds. He was only a few hundred metres away, but there was nothing I could do.

It was well over an hour before he got back on comms, an hour in which he hung spinning in his harness, blood pooling in his extremities.

'Mate, I'm on the move, but I can't stop myself. I'm cramping up, think I've got heatstroke. I need you to stop me.'

I could see him zipping downwards, far faster now than he should. The only way I could do anything to help was to jump on his rope with all my weight, pulling it taut and slowing its progress through the belay device. Filthy swearing that would make a naval cadet blush echoed around the hot rock cauldron. Fifteen minutes more, and he was close enough that we could shout commands to each other. He was exhausted, dehydrated and in a lot of pain – but finally he managed to swing himself down to join me on the boulder. He was white, and his tongue sounded thick in his mouth.

'Jeez, Aldo, are you OK?' I stammered, as we got him off the ropes.

'No,' he answered honestly, 'completely hit the wall. I'm cramping up.' As if to illustrate the point, he tried to undo his belay device, but his forearms and hands had twisted into useless claws. 'This is beyond pathetic. The Marine boys are going to have a field day when they find out I got smashed by an abseil!'

'Don't be crazy,' I said. 'You've got heatstroke. We need to get you cooled down and hydrated.'

First, though, was the task of getting Khaled and Justin down. Khaled descended screaming, his voice echoing throughout the canyon, warbles of terror like a banshee being drowned in a concert hall. I clasped the radio to my mouth.

'Justin, what's going on? Is he OK? Is there a problem?'

'Yeah, he's fine,' Justin sighed.

'Well, what the hell is he yelling?'

'Well, let's see . . .' pondered Justin as he translated in his head. 'He's saying: "I am the sultan of crazy, the king of madness. Weep before me ye mortals, for I am the canyon king." Or words to that effect.'

Far above me, the speck of Khaled on the ropes was growing larger by the second, the lines jumping and bobbing like the anchor line of a yacht in a gale. His screams to the heavens reverberated around the walls in billowing echoes, the caterwauling of a man embracing insanity in himself.

Khaled's attitude to adventure, and to life in general, is different to that of other top-class adventurers I've met over the years. If I had to use one word to sum them all up, it would be 'control'. Though some may be thoroughly hooked on adrenalin, they must all have their habit under control, or they won't last long. The wingsuit flier or solo climber who just has to keep pushing the limits to get that last hit will eventually push too far, and it doesn't matter how tough you are, no one can punch through rock hurtling towards them at terminal velocity.

Climber Leo Houlding is my friend who has been most honest about his battle with adrenalin addiction. 'One day you get your fix from climbing a big rock route,' he told me, 'but soon that's not enough. Soon you have to climb it without a rope. But then even that becomes banal, and you have to BASE jump from the top. Eventually, even that doesn't give you the hit you need, and you keep pushing the limits further and further, until one day gravity comes calling. You either control your addiction or you die.'

Other adventurers have subtly different forms of the addiction. John Arran is a god among rock climbers, though you would never know it from his demeanour. John gives no sense whatsoever of seeking adrenalin – he is an extremely composed and controlled person. He does things that would be impossible to most people . . . but are so far within his personal abilities that he can make them safe.

Though I am not in the same league as John or Leo, I have done many things that to most people would seem insane. Why dive alongside a giant Nile crocodile, or hand-catch the most venomous snake on earth? When I was a young man, it was unquestionably because I loved the thrill of it, because I was an old-school adrenalin junkie. Now it is just the opposite. It's about composure and experience, about truly managing to conquer fear, to grind it under your heel. It's about gaining the strength to operate when others would crumble. To keep your head when all about you are losing theirs.

Khaled is different. Almost 60, he has explored more new canyon systems than possibly anyone else on earth. And yet he cherishes his crazy. He whoops and hollers like a teen on acid dancing on a speaker with their shirt off, fizzing with energy and enthusiasm that it seems will pop his cork off. He is a wild man, full of fun, but with the sense that he could do absolutely anything. When he touches down, his eyes are practically bursting from his head.

Justin, however, had the same hard day as Aldo, and his descent was, if anything, worse. It was pitch-black by the time he started down the line, and by the time he reached me he couldn't physically stop himself descending at all. It was all down to me jumping up on the rope to slow his fall.

When we got away from the ropes both Aldo and Justin collapsed, cramps twisting their bodies into weird shapes. But we couldn't camp here; it was in the fall zone for rocks from above. Our only hope was to push on down in the darkness.

Aldo and Justin opted for just crashing out there, but Khaled was leaping about with enthusiasm. 'No, are you crazy? We have to go on, there is water and flat places to sleep just below, we must press on!' He was up and off.

It was no more than 20 minutes down to the flat, but it was a tense affair. Heatstroke had both Aldo and Justin acting out of character. Every bush that caught Aldo's rucksack was met with kicks, punches and roars, while Justin was so unsteady on his feet that he was in danger of plummeting over every little drop. Both were shouting obscenities at Khaled and me, questioning our judgement, plan and parentage.

'Aldo, listen, mate, you're our medic,' I said. 'You must know that irritability is a sign of advanced heatstroke?'

'Sod off and die,' he shouted.

Dawn brought with it the most curious acoustic. The rocks immediately around us deadened all sound, yet speaking above a certain volume made the vast amphitheatre echo as if you were in a basilica. We were in the wadi, the dry streambed itself. Fear of flash floods usually meant not sleeping in the waterway, but we reasoned that getting up above it at night and exhausted would be even more dangerous. We'd spread our mats around three stepped green pools, which were buzzing with life. Each one was filled with at least a dozen Arabian toads, and four different species of dragonfly flitted around, feasting on smaller flying insects. The ghastly alien shape of water scorpions crawled along the pond bed before popping their bottom snorkel to the surface to breathe. They're ugly brutes, big as my thumb, with raptorial forearms they use to snag their tadpole prey, before puncturing them with their rostrum, a sort of pointy beak, which they use to inject digestive enzymes. Once their food turns to mush, they suck the meat soup up through the same rostrum straw. Nice.

Giant hornets and butterflies landed at the waterside to drink. A wadi racer snake, sleek and speedy, slithered through my fingers and dived into the water. The place was an oasis in miniature. But how? These waterways see rains for hours a year, and then they're flushed through by a torrent that must surely scour every pool of any life. And most of these lifeforms are not specially adapted, certainly not what we would term as 'extremophiles' that specialize in living in the world's most challenging environments.

We started the day quite slowly. Both Justin and Aldo were battered, and Aldo was particularly sore, as he felt he'd shown weakness.

Though what we'd descended yesterday was the biggest drop in the canyon, we still had over a vertical kilometre to descend, and it was going to be a massive day. Particularly because we had packs on our backs that would sink a passenger ferry, and a big pack can wreck everything. All of a sudden, the violent beauty of the cliffs and pinnacles about you is nothing, as all you can think about is the pain in your shoulders, the creak of your hips and knees, and doing everything you can to place each foot carefully so the added weight doesn't topple you. No matter how carefully you try to pack the bag, something hard inevitably ends up pressed into your spine, and it's always heavier at the top than the bottom. Whenever you need something urgently, it involves taking literally everything else out to get to it. You stagger about like an arthritic tortoise and sweat like a jogger in a rubber jumpsuit.

It's a flagellant's way to ruin a nice walk in the desert and it seems to run contrary to modern methods in adventure. Everything in the cutting-edge world of outdoor sports is about getting faster and lighter. Mountaineers who would once have laid siege to a mountain over months using a battalion of porters are now sprinting up them in their trainers. But exploratory canyoning is different. You have to carry double the length of rope of the biggest potential drop. As we'd estimated our drops from satellite imagery, we'd had to factor in a huge margin of error. You have to carry a rock drill, extra batteries and metalwork in case there aren't enough boulders and trees to abseil off. You need enough food and camping kit to be in the field for days on end, and, to top it all off, you need huge amounts of water, just in case you don't find any lying about. It's brutal.

We stumbled down the wadi painfully slowly. The riverbed itself was either smooth and sloping or like a neatly gravelled drive to a Barratt home. It would have been the easiest walk in history, were it not for the boulders. These had tumbled down from the rock faces above us, a limestone Jenga puzzle locked into place by the occasional rampage of the floods. Unlocking the route down is like a bruising crawl through a rocky escape room, always struggling to solve the puzzles at each turn that unlock the next segment of the challenge. A few winds, a scramble and a slide later, and we were at the head of the next abseil. It would drop us into a tight bowl, with overhanging rock walls on all sides.

It is places like this that highlight the genuine exploratory nature of what we're doing. People often ask me how we can be *sure* that we are the first in any of our expeditions. Who's to say if a jungle river wasn't once traversed by a clansman on a hunting trip − even if it was hundreds of years ago? Modern adventurers leave their mark in the written record; if you're going to risk the expense and effort of making the first ascent of a Himalayan peak, you make damn sure to let the Alpine Club know about it. But local people wander all through their world and don't have such a need to stamp their ownership on a place.

All through the canyons of the Middle East there are signs of human habitation going back thousands of years. On the most precipitous slopes are microscopic paths, built up with stacked blocks, dry stone walls and faint trails. A wadi will have boulders scuffed into a polished shine from the passage of thousands of hands and feet over deep time. You'll often happen across goats wandering in places you just don't expect a goat to be able to get to, a sure sign that there is a way into a canyon that isn't immediately apparent. The Omani people can scamper up and down scree slopes that would make an Olympic gymnast dizzy.

But there are some places − like here − where even the mountain goats could never get. Here, you know for a fact that only canyoneers with ropes can travel, and they leave their trace in the form of records, rock cairns, the remnants of slings and rope, and drilled bolts. If none of these things exist, then it must be a first descent. Here, the tatty pieces of thread that anchored Khaled years ago are waiting. If these don't exist in our undropped canyon, we are the first.

We pushed on down Wadi Naqab, abbing off bigger and sketchier dry falls, moving slowly and carefully. The fall was about 80 metres in height, but was not completely dry. It soon turned lush and green. Ferns, shrubs and epiphytes grew over a dense wig of moss; flowers tiny and grand bloomed through the green. It was like abseiling down through one of the great glasshouses at Kew.

Halfway down, we could see that our ropes ended in a deep pond the size of a leisure centre pool. With steep rock walls surrounding the water, and carrying our huge bags, we would most certainly have drowned. The only option was to detour the abseil. Halfway down there was a narrow ledge, covered in a mess of stones and rocks. I kicked as many of them off as possible; the last thing I wanted was for my ropes to dislodge one to fall down on me. Each one plummeted down into the water, causing explosions of spray and booms that echoed round the grotto. I tiptoed along the ledge, traversing out to the side until the water below me was shallow. I then swung down back onto the rock face, and started down again.

It was a high-risk strategy, and it didn't pay off. After five strides, physics stepped in. The rope pinged sideways, and I was fired off on a massive uncontrolled pendulum, crashing me into a rock flake. I spiralled on the line, swearing and cursing. And now I was back over deep water again. My bag splashed down, and started filling up with water. My cursing increased in frequency, volume and creativity.

I tried to reverse the pendulum, pushing one way in order to swing back the other. This did not go well. I was knee-deep in water by now, and starting to feel my tension rising. I couldn't get back up the ropes, I was sinking, my bag was pulling me down. My boots scrabbled on the green slimy wall. What a way to go! I reached for my knife, ready to cut my rucksack straps, cut my rope, and swim for it. And then my feet made purchase. I was standing on a thin ledge, in chest-deep water. Edging my way around the pool, I got into the shallows, and then out onto the warm rock. I secured the ropes so Aldo didn't have to go through the same problems I had. Then I stripped off. There was a hefty chunk out of my knee, and I was going to have a heck of a bruise. But the cooling water was drawing me back in. It turned into probably the finest swim I've ever had.

It was dark when we finally exited the canyon, exhausted. The rest of the team had made their way down, then driven all the way around to the end of the canyon. They'd been waiting for us for hours, and they had no idea where we were or how we were doing. The canyon walls made communication impossible from the second we started, as even satellite phone calls and radio signals did not work inside. Now that the whole team was back together, we chatted about how this was a challenge we'd need to be very aware of as we moved on to the undropped canyon. There would be no one coming to save us if anything went wrong.

It was several days of driving into the mountains to reach Khaled's mystery location, where the undropped canyon could be found. Before we attempted it, though, the reunited team decided to visit one of the villages in the mountains that had suffered worst from the water shortage blighting these lands. Approached through a rock gorge as constricted as a pre-Great Fire of London alleyway, it was like some kind of tableau of an oasis town. The village itself was shaded by green palms, mostly dates.

The only way the village could function was by an ancient system of allocating water. Tiny springs that emerge from the mountains high above are diverted to specific fields and trees based on the falaj system, which allocates water to people during specific hours of the day. The system allows more water for those with more land and with higher seniority in the village. The water itself runs in the most primitive irrigation imaginable; it's dammed and then allowed to run down rocky channels into different places for specified amounts of time.

In the white plaster buildings, behind carved wooden and ornate wrought-iron doorways, we sat with some of the elders and discussed how things were changing in the village. Until the last few years, life had barely altered in several millennia, but now overuse of water resources in other parts of Oman seemed to be wreaking havoc here, where water has always been scarce, but predictable and therefore manageable. It was sad to contemplate the demise of one of the most tenacious cultures on earth, eventually coming to an end because of the profligacy of people who take more than they need.

Two nights later, we found ourselves camping in another deserted village, yet more evidence of a land being desiccated, sucked dry of water. This one sat on a high hilltop; we guessed a gloomy

void to the south must be our canyon. The buildings were composed of stone blocks and looked as though they had been there for
hundreds of years. Despite their crudeness, a couple had television
aerials and electricity cables running to them. There were still
blackened pans hanging from the rafters, walls were sooty-shiny
from their last fires, and the floor of the outhouses were thick with
goat and donkey droppings. It looked as though everyone had just
finished dinner and then upped and left. Outside it was windy; a
desert dust whipped about and stung like gnats. All around the slopes
of the village were acacia trees with thorns as long as my little finger, the sort that'll pierce a sleeping mat; I've even had them pierce
right through the sole of a trainer. The sensible choice would have
been to sleep inside one of the huts, but they creeped me right out
so once more I bedded down under the stars and full moon.

At first light, we stretched and yawned the night out of our
muscles and bones. This time we'd decided to take the full team
into our canyon. Satellite imagery showed an overall drop in the
canyon very similar to that in Wadi Naqab, but over a greater distance. There was nothing to lead us to believe there would be any
drops as extreme as the fireplace in Jebel Shams. So we were going
as a full team, which meant increasing the amount of rope for
safety, and carrying all the camera crew's personal camping equipment. We had no idea if we'd find water, so were carrying six litres
each. I couldn't get my bag off the ground without help, and could
only really stumble downhill under its bulk. If there was much
uphill to do, there would be some serious suffering.

The yomp downhill followed goat trails, hugging the contours of the hills. We didn't commit to getting down into the dry river-bed below us – though it looked inviting – knowing the huge boulders would slow our progress. The overhangs above us trailed scraggly bushes, with little green fruits – capers, the kind you get pickled on pizzas. These were kind of dusty and crunchy, but still tasted like posh salad. Trees sprouted occasional red pomegranates. Broken open, they revealed seedy, pop-in-the-mouth red fruits. It is yet another miracle of the desert that so much grows here in this arid place, and there is so much fruit, bursting with fluids and sugars. Plants don't give these up easily. Trailing on the ground were vines with what looked like small melons growing from them. Gray broke one open and took a bite, then promptly threw up into the dust.

'Don't eat those,' Khaled warned, 'they're poisonous.'

Graham gave him a look that said, 'You could have told me that *before* I tucked in!'

By midday, I was starting to lose heart. So far it had just been a punishing slog downhill in intense heat, but the canyon was all walkable. Everywhere I put my hands or toes on the rocks, the polish was unmistakable. People had been walking this way for generations. We did a couple of abseils to ease our passage down, but a nimble and motivated local could have easily scrambled down. All this effort, and it seemed our undropped canyon was nothing of the sort. What's worse, there were no pools. No water. We had started to suck our resources dry and began making calculations with genuine concern. What happens if we don't find any water by nightfall? Do we carry on? At what point do we start to panic? I found myself down to less than a litre, and there was still none in sight.

Deserts give you a heightened perception of what water means to us as humans. It's one of those things we take for granted in the modern world, where you can just turn a tap and get as much as you want. Here, though, it is everything. As the day progressed, the gorge tightened and steepened, and we found a small pool no bigger than a bathtub filled with dirty green slime. We began pumping it through our water filters, filled our bottles and drank.

As I plunged my bottle into the water, an Arabian toad dived in, swam across and jumped straight onto the bottle, clasping it for dear life. This is what's known as 'amplexus', the love squeeze that a frog embraces his intended with. (Sadly, they often get a bit confused and end up hugging different species of frog, bemused fish, occasionally a human foot. This is the first time I've seen one quite this blind, though!)

We ended a tough day in a boulder gulch, mountains towering on all sides. We stopped at the top of a dry fall that looked as though it must have been 80 metres high. And it dropped into a bowl with overhanging cliff walls on all sides. Impenetrable. Inaccessible. And ours. That night, we bedded down on the rocks, dreaming of our plunge into the unknown. I took from my bag my personal kit, a mat and mosquito net. I'd left my toothbrush and cleaned my teeth with my finger. Cameraman Graham also took his personal kit (which I'd been carrying) out of my bag; a tent, mat, pillow, sleeping bag and a folding camp chair. I forgave him, though, as he'd also packed a bottle of fine single malt.

Pilgrims and poets have always journeyed to the desert in search of . . . something. A desert expedition offers rewards in ways you don't find in other environments. You flog yourself through day-times where the sun seems to be actively trying to flay the skin from your body. You experience thirst and fear like in no other environment. And then the sun starts to set and the light turns into a kaleidoscope of every cliché about colour you can imagine. And then it really starts to get good. Like now. The waning moon cast ghostly white light and flickering shadows over the vast cliff walls. The stars were so clear they were like fairy lights strung up across the black, black cloak of the heavens. And the silence. Quiet that is so ferocious it screams at you. So silent, it's loud.

The next morning, it was my privilege to make the first descent into the unknown. The drop was big, but not overly committing. About halfway down the abseil, a big fat hornet started hovering around my ropes. She disappeared out of sight, then promptly stung me on the neck. For a millisecond I let go of the ropes and slid downwards, before cursing and stopping to rub my swollen neck. Minutes later, I was down on solid rock and surveying my surroundings.

I was in a grey rock bowl . . . well, more like a vase, really, the walls shiny and precipitous around me. At the bottom was a pool the size of a tennis court, and maximum thigh-deep. It was the clearest we'd seen so far, and buzzing with life. A black-headed bulbul paused on the oleander bush beside me and eyed me quizzically. A grey wagtail bobbed up and down on the other side of the pool from me. A huge hawker dragonfly in azure blue colours darted about the surface, chased away by territorial smaller red dragonflies. Then the air was full of small birds. Martens swooped in to drink at the pool while still on the wing, snatching a beakful of water as they sallied in sorties over the surface. There was no polish to the rocks, no cairns, no remnants of gear left from ancient abseils. No human had ever been here before. Just me.

All I could think was, *I wish Helen could see this*. I sat down at the waterside, looked down at the bugs doing their thing, wondered at the beauty of it all, and felt very, very lonely indeed. There was an overpowering physical need to hold my little boy in my arms, feel his burbling breath and his snot plastered onto my face. He would have bounced up and down with glee at the birds, and splashed like a wild thing in the water. Just as it comes to the first ascender of a mountain to give that peak a name, that honour also goes to the first descender of a canyon drop.

'Logan's Pool', I decided. It seemed unlikely I'd ever get the chance to bring my son here, but who knows?

The rest of the team were down with me soon after. Rosie was also quite quiet, and I could tell she was thinking about her little ones and how far she was from them. Aldo sat on his own, staring into the water. He had his lost ones and demons as much as any of us. We all refilled our water, silent in our own little worlds.

And then Khaled came down, shouting and screaming with excitement.

'We are here!' he exalted. 'We shall call it Steve's Drop!'

I didn't correct him, even if Steve's Drop sounds more like a hastily fashioned camp toilet than a place as perfect as this.

'You know what?' Parker, our sound recordist, said soon after. 'I was lying in bed last night, aching, sore from my bag, hungry, thirsty. And I thought to myself, what would the 15-year-old me think if he could see me right now? He'd lose his shit! Imagine, I get to come on expeditions, go places no one ever went to before, and do it all with my pals. It just doesn't get any better than this.'

I turned to him and smiled, placing a hand on his shoulder. 'Not pals, Parks . . . colleagues. Don't get ahead of yourself.'

Many hours and abseils later, I was taking the lead, with Aldo and Justin fixing ropes, and Khaled dashing along with the enthusiasm of a King Charles Spaniel and the sure-footed nimbleness of a man half his age. We dropped into yet another new bowl, with a deep, dark green pool filled with vegetation. Khaled skipped over the rocks and a small tree at the base, then shouted, as if he had been stung by a hornet, 'AAAAAGGGHHH! Viper!'

'Oh God!' I leapt to my feet. 'Have you been bitten?'

'No, no,' he chided. 'Here, beneath the ropes, it's a viper!'

My heart beat faster. Snakes are my obsession, and have been since I was a child. They are also the bane of my adult life. I spend hours and hours of every day that I am out in the field looking for snakes, barely ever finding them. The everyman expectation that huge venomous reptiles lurk in every tree and under every stone in wild places is the greatest curse for me. Because it isn't true. Snakes are hard to find, and rare in most environments. It's not uncommon for me to spend weeks searching and find nothing at all. Every single one is still a triumph for me even now, and there was one here I was desperate to see. The saw-scaled viper was the first venomous snake I had ever caught, over 20 years ago, and it was special for many reasons.

My mantra is that wild animals don't mean us humans any harm, and that their danger to us is massively overstated. All the world's sharks kill fewer than ten people a year around the world; you're more likely to be killed taking a selfie, being struck by lightning or crushed by a falling soft-drinks machine. The exception that proves the rule is the saw-scaled viper. It is the most dangerous snake in the world. Occurring across northern Africa and through the Middle East into Asia, these vipers are ambush hunters, lying in wait in good hunting positions – sometimes for days on end. They strike out at anything warm that passes nearby. Their camouflage is superb, and it is easy to miss them, even to step right on them. Human fatalities from saw-scaled vipers are highest in rural Sri Lanka, Bangladesh and India, where many people work barefoot in the fields and are loath to even try to get to medical help. This little snake, despite not being in the Top 20 in terms of the toxicity of its venom, may be responsible for 20,000 human deaths a year, though that number is impossible to prove. The world's most venomous snake (the inland taipan) has never caused a recorded human death.

I looked under the rock Khaled was gesturing to and caught my breath. Beneath was not one but two saw-scaled vipers, sandy in colour, with dusty scarlet diamonds running down their backs. One was larger and fatter, thus clearly the female. As I leant in, they began the threat display that gives them their name, sliding their coils sideways against each other, the heavily keeled scales rasping together to create a sound much like the static on an old-fashioned radio or bacon frying in a pan. It's hypothesized that this threat mechanism evolved to replace hissing in these desert snakes, because hissing evaporates water from the mouth, and desert snakes can't afford to lose fluids. It is a chilling sound. Just as the rattle of a rattlesnake will scare away a calf that has never seen or heard the sound before, the fizz of a saw-scaled viper will repel even ingénue creatures. I gently picked up the larger female to examine her. The bulges in her belly suggested she was feeding well on the Arabian toads in the pond before her. She struck once towards me, exposing her hinged fangs. Taking care to make sure that everyone coming down the ropes knew where she was, I carefully returned her alongside her mate.

We came to a portion of river valley that had clearly seen a relatively recent flood. The floor was thick mud, cracked into a patchwork like the scales of a dusty dragon. There was a really unusual consolidated rock packing in the riverbed – rounded stones packed into a matrix of what looked almost like concrete. As we progressed downwards, all the most prominent boulders held a little jumbled pile of fruit stones and seeds in purple dried goo. Aldo stooped and took it between his fingers, smelling it.

'Fruity!'

'It's ripe gorgonberry,' I said. He took a taste and spat it out.

'Either that, or the droppings of the rare Blanford's fox.'

Above this the pinnacles were composed of packed layers of different rock types: ochre, ebony, chestnut and cream colours, stacked in bedding planes like a *mille-feuille*. Puffy clouds billowed and rolled across the cliffs, casting fast-moving shadows. Some of these clouds started to build and gather, getting steadily higher, and forming the classic anvil shape of the thunder cloud. Abdul Hamid started to quicken his pace, almost sprinting ahead of us.

'What's your hurry, Abdul Hamid?' shouted out Justin. 'There's no "I" in team!'

'Yes, but there is a "U" in flood,' he replied.

'No, there isn't!'

'What would you know? You barely speak your own language.'

The banter between the two was tinged with seriousness.

Thunder clouds could mean rain, and rain would mean a flash flood. And we were in a tight canyon, with no escape. The unsaid truth was that we still had several hours to go, and if a flood ripped through here, we would all die. Cameras were quietly packed into bags, feet started to whirr across the rocks, tired muscles lost their aches. We slid or jumped down small drops where we might have otherwise rigged ropes. Eyes narrowed, conversation stopped. Everything took on an air of complete seriousness.

It was several tense hours before the last abseil, which dropped us down into a final boulder field. Now we could see a junction with the main canyon ahead. The rain had yet to fall. The walls above us lessened in steepness so we would be able to reach safety if a flood thundered down. And, as we hit the last wadi, there was a sight I would never have believed possible: water. Not the slimy puddles we'd been syphoning our water from, but fast-flowing, clear, deep water. It thronged through canyon walls as tight and steep as a city alleyway, so close you could reach your arms out and touch rock on either side. I knew that above us was nothing but mountains and empty desert, so I plunged my whole head into the water and guzzled it down, drinking till I felt I would burst. It was cold! Chilled, pure water in the middle of the desert! We all laughed like residents of *The Rubber Room*, releasing tension and pure joy. There was still a long way to go to get back to civilization; the following day would be a big slog, we wouldn't get out of the canyon till late at night . . . but for now, splashing around in the dying light, we were canyon kings, sultans of crazy.

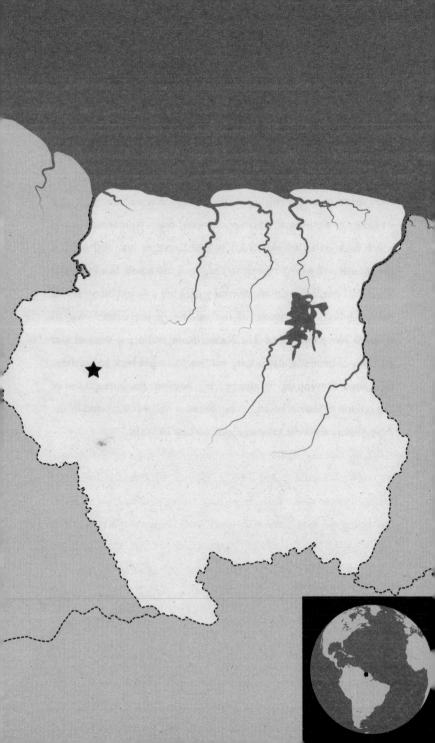

6. GHOST RIVER

SURINAME

Scrabbling round on the rocky bottom, my fingers gain purchase on the lip
of a rock. I drag myself down, pulling my chest onto the riverbed, trying to
make my body smaller, in the hope that somehow my delectable extremities
will become less evident. The current carries away the flow-borne sediment,
and the clouds clear for a few seconds. It has the look of a Hollywood horror
feeding frenzy. Hundreds of piranhas of varying species and sizes swarm in
the gloom, moving with erratic, jerky, staccato twists and terrifying speed.
The sound is overwhelming; through the water the click and clack of the
sharpest teeth in nature duet with ripping, tearing flesh and thrashing tails
and fins. Blood-red eyes from the largest piranhas on the planet turn their
ghastly gaze through my mask, mere millimetres away. Their underbite
reveals interlocking scalpel-sharp teeth. I lie motionless, willing myself to
melt into the riverbed, clenching my toes together into unappetizing clumps,
focusing on one of the most horrific predatory spectacles in nature.

We'd been flying for nearly an hour over an unending duvet of tightly bunched trees on ridgelines, riversides and broad expanses of flat forest. Occasional bright yellow blooms in amongst the green were trees in full yellow flower, utterly breathtaking, even from 10,000 feet up.

The most spectacular rainforests in Borneo, Sumatra and Madagascar are now a paltry memory of what they once were; dustbowl plains where once were seemingly infinite forests. Here, though, was different. Suriname is the greenest country in the world, with at least 90 per cent of its land mass covered in trees. It has a lower rate of deforestation than any other jungle nation, and a population of 500,000 in a country that's almost as big as the UK. It's just empty forest, and in a unique position on the Caribbean coast of the South American continent. This is the most biodiverse place on earth, with more species of plants and animals than anywhere else. It's also one of the least explored.

When I first started looking into expeditions into the interior of Suriname in the late 1990s, I found rivers that had not even been mapped, and mountains that had never been climbed. It was a genuine shock to go back to my planning 20 years later and find nothing had changed. There was a dizzying amount of new things to discover here, a whole new world of wonder just waiting for us to drop in.

Our main issue was going to be accessing any of these untrammelled regions. They had remained unexplored for a reason: they were all staggeringly remote. In the 1960s, a project known as

Operation Grasshopper had sent teams into the interior looking for precious metals, oil and gas. Some of the prospectors never returned. Those who did came back battered, bruised and convinced that it simply wasn't worth the effort. Any airstrips they had cleared had been reclaimed by the forest. We were going to have to start from scratch.

Most of my expeditions are in tropical rainforest, and I've learned my lessons the hard way. Just hacking off blindly into unmapped jungle is one of the least productive enterprises imaginable. Navigation is tricky and progress can often be measured only in hundreds of metres a day. It's hard manual labour, with little or no reward. Every animal scarpers at the sound of your machetes.

The best chance of progress is using rivers. The waterways are the superhighways of the forest. On the big rivers you can cover excellent ground in a variety of different powered or human-driven craft. The creeks and gullies offer stunning scenery. There's the promise of a wild encounter round every corner, and the oxbow lakes and rapids are home to crocodiles, frogs, snakes and colourful birds to delight any naturalist. If you want to explore in the rainforest, you want to be on a river. Of course we're not the first people to realize this. As wild as Suriname might be, it's been a fascination for Dutch explorers for over a hundred years. The big rivers on the maps have mostly been surveyed by military teams and fortune seekers.

However, satellite imagery has started to show rivers and creeks in Suriname that are hundreds of kilometres long and have never seen an explorer's boot. Many of these creeks are too shallow, rocky and fast-flowing for a motor boat to travel up. The only course of action then is to descend them. So ideally you would trek through the forests into the mountains where they originate, carrying your boats with you. This is an exhausting and time-consuming enterprise, which would have taken most of our year. The alternative would be to use a helicopter to drop you upriver. And to do that, you need a dependable landing site . . . in the middle of all the endless broccoli. We spent many days and weeks on Google Earth studying every millimetre of the various possible creeks, looking for a weakness, seeking the place that a waterway might give up its secrets.

My secret weapon was a half-Dutch, half-Surinamese adventurer by the name of Michel Boeijen. Michel and I had met way back in the late nineties, when I was working for *National Geographic* (with the rather grand title of 'Adventurer in Residence'). He had instantly impressed me with his boyish enthusiasm for the jungles of Suriname. Back then there was no satellite imagery, so we pored over maps together, making plans for grand missions into the unknown heart of the country, little knowing it would take 20 years to make our dreams a reality. Michel had brought in two locals to aid our exploration of the dark heart of Suriname. Keiran was of Amerindian descent, and effortlessly knowledgeable about everything that lived and grew in these forests. Ile had been with Michel on many of his former exploratory expeditions, and was from the village of Djumu, where our journey had to begin.

Eventually Michel found something: a glint of gold on one of the older satellite images, telling of a creek running down from the mountains. He asked if I had any ideas for a name.

'Well, it has no name, it doesn't occur on any maps, it's a phantom, a ghost . . . can we call it the ghost river? Ile, what's the local name for ghost river?'

'YiYi?' Ile pronounced. Michel clapped his hands in delight, and with that, our creek was christened.

Pre-colonial rule, Suriname was thinly populated by what are now known as Amerindians – the Caribs, Arawak and Sipaliwini, many of whom had been here for as long as 10,000 years. They practised nomadic agriculture, growing potatoes, cassava, maize, peanuts, tobacco, cacao and tomatoes, creating small-scale forest gardens, which they cultivated until the thin soil ran out of juice, then they moved on. At the time the first Europeans arrived in Suriname, the estimated population of the interior was just 18,000. During the era of slavery, the wilderness of the interior was an inevitable attraction to runaways, who slipped their shackles and made for freedom. Plantation owners sent hunters in after them, but with limited success. For their part, once established in small riverside towns, the slaves sent raids of their own into the plantations, taking women, food, weapons and supplies. Soon the folly of trying to fend off their advances left the plantation owners to seek settlements. Slaves became free in Suriname a century before they did elsewhere in the Caribbean. Their descendants still live in the interior today, living lives that still owe much to their West African heritage. These peoples are known as Maroons, and in some cases are still nervous of outsiders.

Within just a few days of our arrival in Suriname, we found ourselves making camp on an island in the middle of the vast Brokopondo Reservoir. Dammed for hydro-electric power way back in the 1960s, the lake contains the sunken relics of the villages that were flooded, now inhabited by crocodiles and giant sawfish. The great forest trees that had been swamped now stood proud from the water, without leaves or branches – just bare totem poles, their dark reflections shimmering on the mirrored water. We'd cross the lake by motorboat to begin our penetration into the interior.

Rarely have I felt so far from home so quickly. At night I battled the furious humidity by lying in the shallows, with little fish nibbling my freckles. Lightning strikes ripped across the sky in the distance, reflected perfectly in the glassy black of the lake. In normal circumstances I'd have been swept with waves of pure ecstasy on seeing something so beautiful, knowing how intensely privileged I was. But now it was tinged with melancholy. I'd had ten days at home with Helen and Logan after the desert, which was long enough to see him change from a wailing baby sack that did nothing other than eat and sleep to a tiny person. He broke into brilliant infectious smiles, recognized me, and already had a firm grip on my heartstrings. This Suriname expedition was to be five and a half weeks, with no opportunity to contact the outside world. That seemed like a long time.

Over the following two days we crossed the lake and made our way upriver towards the village of Djumu, the last frontier town before the vast emptiness of the interior. The village was much like landing on the banks of the Congo, with wooden thatched houses and black African faces. I've travelled through most of South America, and this is certainly unique. These people identify themselves as descendants of escaped slaves. It's been said that anthropologists wanting to study West African culture should come here to Suriname as the culture remains more true to its origins than in West Africa itself.

Around the town there were copious symbols – offerings in near human form made to the gods. There were also offerings of alcohol and rice left out for the spirits that exist in everything from the trees and rocks to all the important kinds of animal. In order to gain permission to continue with our journey, we needed to first get permission from the Basia, a kind of shaman, who shares the duties of protecting the village's heritage with the 'Grand Man', or village chief. If by any chance the Basia said no, we would have no choice but to turn tail and head for home.

The night was spent in Ile's house, as we waited for permission to speak to the Basia. Pretty soon the local kids found us, and spent an hour or so getting steadily closer and closer until we were all playing football together around the fruit trees and palms. When we finally collapsed, exhausted and dripping sweat, our new friends went through our bags to take any tiny tidbit that could be offered as a present.

Initially they were nervous of the cameras – they still had an inbuilt distrust of them, and believed having a photo taken removed a little of their spirit. However, after one of the bolder girls posed for a photo and saw the results on the screen, everyone was queuing up to have their picture taken in a giggling morass of enthusiasm. We could only hope to have half the same welcome when we met the headman.

The following morning we were received in a thatched-roof wooden house. The dark interior was lined with a variety of weapons, from ancient muskets to handmade shotguns held together with red tape. Crudely carved wooden masks and mannequins battled for space on the walls with colourful umbrellas and cheap wall clocks. The Basia himself was a rheumy-eyed, dark-skinned man in perhaps his seventies, tall and rangy, with close-cropped white hair.

For the first half an hour, a tense conversation took place between Ile and the Basia in their local dialect. None of us had any idea what was happening. Ile appeared to be super respectful, even obsequious. It didn't look good. Aldo, Michel and I sat with bated breath, nodding earnestly in agreement at things we did not understand, trying our best to look like the kind of people he could trust. Eventually their discourse thinned enough for Ile to translate some of the dialogue into Dutch, and then Michel translated it to us in English.

Eventually we had the Basia's pronouncements relayed to us. Above the village were lands that the people of Djumu never visited, as they believed they were home to the discontented ghosts of their ancestors. Several hundred years ago, slave drivers had travelled inland to try to recover their escaped slaves. They'd brought with them heavily armed mercenaries, and when they arrived, they unleashed fury on the locals. The Basia recalled this as the time the rivers ran red with blood, so much so that they named their waterway Blood Creek. Despite the ferocity of these attacks, the people of Djumu refused to yield, and eventually the slave owners gave up. None of this sounded good, or seemed to bode well for the Basia's answer.

'So what does he think?' I asked. 'Will he give us his permission to travel into the interior?'

Michel translated this to Ile, for him to translate again for the Basia, but instead Ile just replied instantly in Dutch. Michel shook his head. 'Ile says he said yes to that straight away, he just wanted to tell us the story.'

We all breathed out. I tried my best to say thank you in the local dialect. 'Gang an di fee,' I tried.

'You should all stay a little longer,' the Basia told Ile. 'Give him a chance to learn how to speak our language a bit better.'

A short hop in a light aircraft took us to an airfield in the forest that had been kept open over the years since Operation Grasshopper. Our airfield bore the memory of that failed project in the form of an ancient tractor that had been brought out here piece by piece, and an airplane, crashed in the forest near the runway, crawling with tarantulas scampering inside the fuselage.

Green palm tanagers hopped around the scrub, eyeing us with

keen black eyes. A great black hawk loomed upright and straight-backed at the riverside. Exuberant great kiskadee darted about in their airborne pursuit of bugs, with their call that translates into local tongues as 'I see you well.' Hordes of falcon-like black caracara are a good omen said to be a sign that capybara and tapir are near, as caracara perch on them and pluck parasites from their fur.

We'd deliberated about craft to use on our unmapped river and had decided on inflatable canoes, which were perfect for carrying the loads we'd need. We had two long boring days where we took apart all of our equipment and tried to pack it all into our inflatable boats. It was like some bonkers game of Tetris. None of us had any more than a small personal bag for the entire journey, but add in an array of cameras (just in case we were to round a corner and see a jaguar in front of us), two generators and fuel, and every boat was soon loaded with 60–100 kilograms of equipment. This was to end up being the most challenging element of the mission.

On the morning of our third day, a tiny white Robinson heli-copter landed at our airstrip. It had no doors, just a single sheet of plexiglass in front of the pilot. It looked like a home-made Mec-cano chopper. With quizzical sidelong glances, we started to prepare to lift all of the crew and equipment over the trees and up towards the source of our phantom waterway. Aldo and I looked at each other ruefully. Robinsons are great for being low-maintenance and manoeuvrable. However, their payload is not big, and worse, they have no back-up to their engine. If you get a malfunction of any kind, you go down. As we were in some of the most remote and impenetrable forest left on planet earth, we would just have to hope we had a cautious pilot.

We did not. As soon as the rotors engaged, the heli ripped forward, then the pilot yanked the stick, flipping us up over the trees, and dropping us down to the main river. He then proceeded to fly so close to the water it felt like we were in one of those Everglades airboats. The water rushed past us just a couple of metres below our feet. We all held our breath. No one wanted to say a word that might distract our pilot; if he had so much as sneezed, we'd have caught a skid and cartwheeled into the drink, ending up as food for the crocs.

Soon a green wall loomed ahead. The pilot pulled back on the stick and the nose headed for the heavens. We whooped like teens on a roller coaster, and then gasped. Beyond the emerald barrier was the spine of a mountain range, lying like a slumbering armoured dragon, covered with cauliflower florets, asparagus spears and broccoli heads. It was stupendously beautiful, but the Jungle Jim in me looked at it with utter dread.

We've done dozens of expeditions in jungle like this, and it's been hell. On foot, you manage to cover no more than four or five hundred metres in a day, and even when you have GPS, getting lost is as good as a certainty. The rivers, though, are our saviour, the lifeblood of the forest. Even the wild animals use them as a means of getting around, because travelling along the forest floor would be like dragging yourself through a web of barbed wire. Using our inflatable boats, however, we might be able to cover as much as ten kilometres in a day.

Our little heli flew for half an hour over pure green paradise. Scarlet macaws took off from the treetops just metres from our feet, glossy ibis flapped down the riverways. And then suddenly out of the green was a glint of gold sunshine on water. It was our river – the YiYi! Our pilot banked to the right, and we started to track up the river course towards the nearby hills, presumably the source. What we needed was a decent open area to make our landing zone. Michel had spent days poring over the satellite images, and had a few possibles that might be open sandy banks. But the satellite images could have been shot at any time of year, and the waterways are constantly changing. Before we reached it, we didn't even know if the YiYi would have running water throughout the year. Or the beaches could be under six feet of water.

The river beneath us snaked gold and brown through impossibly dense vegetation. There were numerous logfalls, where whole trees had collapsed right across the flow. Michel and I made eye contact as we saw them – they'd inevitably be a total grind to manoeuvre around.

'There!' he shouted, pointing.

'It's a beach!' I yelled back. 'Beach' was maybe a stretch. It was more a muddy sandbar the size of a squash court. Trees hung down on all sides, creating a veritable stadium, with our landing zone in the centre. It was going to take some bold flying to land us in the middle of it. Luckily, 'Bold' was our pilot's middle name (either that, or 'Reckless').

He whipped around, shaving the foliage off a bunch of

waterside trees with the tail rotor, and edged forward to drop a skid to the sand. There were bursts of sand, then crashes and splashes as four huge iguanas sprinted from their sandy basking spot and leapt for the water. Michel and I were already out of our seat belts, perched on the skids. We jumped down with machetes in our hands and dashed around, blindly hacking at anything and everything that might entangle our propellers, while the pilot hovered precariously just off the deck. Ten minutes of furious hacking later and the heli dropped, offloading the first of our loads. Just seconds later, and the sound of the rotors disappeared off into the distance, leaving Michel and I alone. The reaction was an instinctive roar of elation from both of us, and we grabbed each other in a massive hug. More than 20 years of plotting and preparation, and we were finally here. Alone. In the jungle. On a river that didn't exist.

It was an hour before the heli returned again, which Michel and I spent wandering our little sandbank in a daze. It was as if we'd teleported into a Tarzan film. The skies were filled with screeching parrots – red and green macaws flew overhead in Spitfire squadrons, and white-fronted toucans wiped their preposterous beaks on branches high above us.

'Steve! Come and look at this!' Michel shouted, enthusiastic as a seven-year-old. At the water's edge in ankle-deep cola-brown water were four gigantic wolf fish, or Aimara, each one as long as my arm. Even after the heli madness they didn't seem remotely bothered. Thousands of sand digger wasps in stripy rugby jerseys churned away at the sandbanks like dogs digging for a bone. There were resonant sounds like woodpeckers, but they actually turned out to be the staccato calls of giant bullfrogs.

When the heli returned, it was not with people, but with a net dangling beneath it with our boats. Michel and I got to work, frantically pumping them up. The adrenalin drove me to jump up and down on the pump and have them all done before the heli came back. The temperature was in the high 30s, the humidity close to 100 per cent, the sky cloudless. And we'd decimated everything even approaching shade on our sandbank. It was ferocious. Already drenched in sweat, I slid into the water, giving our wolf fish a wide berth, and waited for the inbound mosquito drone of the heli.

The monotone green of the waterside vegetation was brought to life by clouds of yellow butterflies, usually known as sulphurs. All of the ones gathering at the riverside were males, and most had only recently emerged from the pupa. They were doing something called mud-puddling, in which they gather the minerals they need for reproduction, pumping the water out of their anus. These butterflies are migratory; in Mexico we encountered clouds of them, like green-yellow blizzards across the roads.

It was six hours before the whole crew and all of our equipment was assembled in the sand, and we could take stock. The general reaction was of utter jaw-dropped awe. Anna, our second camera and producer, had never been in jungle like this before, and the experience was overwhelming. Just our little family, alone in a place where no human had ever been before. I tried to chat to her about how it felt to be here, but she was hesitant, unsure of her words. I realized she was close to tears. I knew exactly how she was feeling. For all the remarkable things we'd done in this bonkers year, I think this day was the most special.

The time it had taken us to get here meant it was already well into the early evening, and as desperate as we were to get started, it made best sense to just set up camp here. It was actually a pretty good site; as the sun started to lose its heat the lack of shade ceased to be an issue, and it was as close to perfection as you could ever ask for. In fact, the only downside was that there was nowhere at all to go when nature called. We decided to dig a little hole in the downstream part of the sandbank, and placed an opened umbrella in front to act as a modesty screen. 'Poobrella' was to become a feature in almost every conversation for the rest of the trip.*

---

* Going for a poo in the wilds can be the greatest, or worst, moment of every day. It is something we all give an inordinate amount of thinking time and planning to. And that is for good reason. Best-case scenario: you wake up, have a strong coffee, feel the motion in the ocean and stroll in the sun to a scenic spot, newspaper under your arm, for a poo with a view. At such moments you feel as if you are winning at life. Worst-case scenario: you get up in a howling Arctic gale, break camp, put on your drysuit, lifejacket, etc., get into the kayak and on the water, and then and only then does your tummy let you know there is urgent business to be done. Slithering in the snow, baring your bottom to an ice hurricane, probably covering yourself in unmentionable goo, while the rest of the team sit around waiting . . . these are moments when you think a career in accountancy might have been a better option. Anyone who is new to expeditions will be astounded at how comfortable we become with topics of the toilet, and how much we talk about it. I did my expedition leader training in the Himalayas with the Indian army, and we

As stunning as our patch of virgin forest was, not everyone was happy. Graham had set up his tripod and massive wildlife lens, and was busy knocking off shots of kingfishers, parrots and toucans. Every time he thought he was done, one of us would leap up and call his attention to something else.

'Is this the best jungle you've ever seen, Gray?' I asked.

'Well, it's alright, you know,' he grumbled, 'but I made myself a cup of coffee half an hour ago and still haven't drunk it.'

Later on, as we were sitting round the campfire, a great green iguana wandered nonchalantly through camp, practically walking over our legs, to take up a proud position standing on the prow of one of our boats, like he was one of those carved mermaid mascots pirates used to have at their bow.

Keiran, our local friend, had to physically hold himself back to avoid jumping on our new mascot.

'But iguana taste so good!' he complained.

'Let me guess,' said Aldo. 'Like chicken, right?'

'Of course!' Keiran responded. 'It's jungle chicken!'

'And what else would you like to eat here?' Aldo asked.

'Well, the caiman, of course,' Keiran nodded.

'Like chicken?' Aldo asked.

'Yes! And the macaw too, they're just like chicken.'

---

had a designated time every morning when the whole squad would squat in a line at a trench we'd just dug, passing the loo roll along the line. That shit makes you pretty relaxed about . . . well, shit.

'I guess for you it's more that chicken tastes like macaw, though, right?' I asked.

'I don't know,' Keiran responded. 'I've never eaten chicken.'

Though we managed to keep Keiran and Ile from killing any of the bigger wildlife, there was no way we could stop them fishing. And despite having lived in the jungle their entire lives, neither of them had ever seen anywhere like it. You could wade into the rivers and catch the wolf fish with your bare hands.

Eventually, after they had a couple of monsters, we convinced them to stop. Keiran handed me the biggest of them. It must have weighed ten kilos, and inside its broad trouty mouth were teeth that looked more like they should have been in the gob of an actual wolf! It was no surprise, then, that local people actually feared these ambush predators more than they feared the piranhas, crocs or electric eels. Their cone-shaped teeth didn't bite out clean wounds like piranha, but cut in like a croc, and then they spun their whole body. They can weigh up to 40 kilograms, so the wounds they inflict can be horrific. Wolf fish will lash out at anything tempting they see in the water, and while that's usually a smaller fish, it could easily be someone's toe. We resolved not to swim naked for the duration of the trip.

Next morning, we were all up with the dawn. The insides of our tents had got so hot so quickly that we all felt as if we'd been flash-fried. Then began the drill of taking down tents, packing all of our stuff, and lashing everything into the inside of our canoes so that it wouldn't shift and slide to maintain as near-perfect trim as

possible. The slightest tilt or lean to the boat from improper load-
ing would become a big issue over ten-hour paddling days. The
packing and unpacking of boats took up at least three hours of
every day, and became just another part of the monotony of jungle
routine. Aldo, of course, loved it.

It was no more than a couple of hundred metres downstream
before we pulled up to the first spot where trees had fallen across the
river. Michel, Keiran, Ile and I waded in up to our chests and started
hacking with our machetes to clear a way through the obvious line.
The water was muddy brown and you couldn't see anything. Yet we
were all certain that it was full of wolf fish, piranhas, stingrays and
much else besides. At logjams like this, their concentration would be
at its most dense. With every thrash of my machete, I was secretly
clenching my toes together to prevent them looking like bait.

The day was hard. Harder than we could possibly have expected.
The logjams came thick and fast, and our progress was seriously
slow. After about eleven in the morning the heat was just searing, and there was no escape. I wrapped my Omani scarf round my
head and neck to cover myself up. It seemed to me that I looked
like Lawrence of Arabia, windswept and heroic . . . until Anna told
me I looked like Miss Marple.

Our wildlife sightings also started to dry up. In the furious
furnace of midday, everything hid away in the shade. We had a
few monkey sightings, and then a bunch of manure in cricket
ball-sized lumps floated down past us. Tapir dung! I reached down
and took a handful from the water; it was fresh. Tapir are incredibly

important animals because of this stuff; they feed on fruit, and the seeds are passed out through their droppings and thus dispersed, spreading the forest tree species far and wide. It's believed some species need to pass through the digestive system of a tapir just to germinate.

A couple of times we pulled round a corner to an explosive splash and a rumble through the jungle as a tapir legged it off into the green. Then we paddled up to a small green cavern at the riverside to see a big female tapir just emerging from the water, her flanks slick and dripping like a horse that's just had a dip. She twisted her bendy nose and turned to eye us with utter incomprehension, before shaking herself down and striding into the bush.

That night, after ten long hours being shallow-fried in our blow-up boats, we gave up on the search for a perfect camp and went for the first place that would work. We wanted nice flat rocks out in the stream, or another beach like the previous night. What we got was a patch of dingy dark forest, but it was at least relatively flat for our tents. The light had faded to near total darkness by five o'clock in the evening, and we sat around the stoves in silence to force down our dehydrated rations.

After several thousand calories of stodge, I decided to head out into the gloom to see if we could find any wildlife. Almost imme-diately I realized we were not going to get lucky. The floor of the jungle was alive, moving with a dark undulating wave. When I looked closer, I could pick out the bodies of hundreds of thousands of army ants. The only thing that allowed us to walk here at all was an ankle-deep stream alongside them. While we stayed in the water, we didn't get bitten. Otherwise, a feeding raid like this would swarm over us and we'd have to dive into the main river to escape. When a column of army ants comes into a native village, they don't even try to dissuade them. Instead, the villagers move out. When they return, the ants will be gone, and they'll have taken with them all the cockroaches, spiders and scorpions, providing a natural spring-cleaning service.

Further up the stream, we made a peculiar find. A huge tree with a hollow trunk swarmed with ants, and also what looked like a couple of bucket-loads of brown rice dumped on the trunk and roots. The ants were going crazy, carrying off the little grains and setting up trails into the undergrowth, with other ants clasping onto each other, forming living bridges with their bodies. These grains of rice were actually the larval casings of their young. They would have been held inside the tree in a nest, protected by the soldiers. My guess was that something – perhaps an anteater like a tamandua – had come in and taken a feed from the protein-rich larvae. Maybe the bites of the ants got too much, or it heard us approaching and did a runner. Now the ants were in confusion, and charging everywhere to try to save their babies, and find a new place to bivouac.

Yet this was not the most special find of the night. In the same shallow stream that was saving our ankles from being munched, I saw the bizarre shape of a Suriname toad. This is surely one of the world's oddest creatures, looking like a normal frog that's been run over. It's as flat as a crumpet, and you could mistake it for a brown leaf until it starts using its star-shaped fingers to swim clumsily. It has no teeth or tongue, with tiny googly eyes on top of its head, but this is nothing compared with its method of reproduction. When the male and female get down to it, she flicks the fertilized eggs up onto her back, where they start to sink into the skin. They develop into tadpoles, embedded in their mum's back, like some weird thrashing honeycomb.

The spiders here were also dramatic. Skeleton tarantulas sat waiting in the mouth of their burrows for prey to scurry past, bright neon markings down their legs giving them the look of a Halloween human in a Lycra skeleton suit. Wandering spiders may not look quite as fearsome, but they are the world's most dangerous, with a venom which can lead to endless and painful priapisms in men, and then total loss of all sexual function. And they were everywhere, scampering into your boots and hammock whenever you were not looking. Then I saw a bizarre all-white spider that I'd never seen before hanging over the water. On closer inspection it was actually a normal raft spider that had been infected with the cordyceps fungus. It was a real horror show; the fungal spores get ingested by invertebrates, and then mature and erupt from their bodies like some kind of parasitic alien. Our white spider was decorated with ivory spines and baubles: the fruiting bodies of the invasive fungus.

The YiYi river had an ever-changing character. Sometimes it was narrow and clogged with trees. Other times it opened out into such a wide, broad waterway that it was hard to believe it had stayed unknown so long. By day three, though, we were beset by constant rapids; and I'm not talking dramatic whitewater cascades with adrenalin-filled sloshes down them. These were places where the flow was shallow, bouncing over rocks and vegetation, and was generally too difficult to paddle through, so you'd have to get out of the boat and drag it. As you did so, the water would inevitably erupt as a nearby croc took fright.

All the rocks that are exposed above the water were covered with thin spines. These germinate as soon as the water flows over them and turn into fronds that look like aquatic cabbage or lettuce, and foot-high stems topped with purple flowers. These are *Pacu wieri*, named for the pacu fish (which kind of looks like a piranha, but with oddly human teeth) that loves feeding on their seeds. Underwater was a veritable salad garden of plants, like paddling through the fresh veg section in your local supermarket.

My boat seemed to have been possessed, dragged towards every single rock and branch as if by some magnetic force. With every outdoor sport where momentum is a factor, the number one rule is to focus on the gap between obstacles you want to avoid. If you stare at a boulder you want to miss, you will be dragged into its tractor beam. However, on this river it didn't matter where I focused; I'd still end up wrapped round a sunken tree. I was spending more time out of the boat wading and dragging it over rocks than paddling. It was utterly exhausting, and I was just begging for camp to come so I could crash out and fill myself up with calories.

Another omnipresent creature as we progressed was the brown long-nosed bat. Sometimes the common names of animals are very much a case of 'say what you see'. After all, it's brown, it has a long nose. And, it's a bat. Almost every bough that overhung the water was home to a roost of them, from four or five up to 20 or more. They stuck on vertical wood or stone, head down, noses pointed upwards, and as you paddled nearby, they took flight as one in a flouncy panicked ballet. They then came to rest again on another suitable roost, sticking there like a damp cowpat thrown at a wall.

That night I came to a piece of flat jungle, which was clearly a practical and logistical winner, but dark and gloomy. I told the others to hang fire, and recced half a kilometre further on, where there was a huge flat rock in the middle of the river. There were paradise-perfect pitching spots for five of us, but the other four would have had to slum it in the forest, and wade through waist-deep, fast-flowing water to get out to us. It was a tough call, but I opted for caution, not splitting the team, and everyone being miserable together in dark forest. I could hear them grumbling, imagining themselves sitting in the sunshine out on the rocks, drying their gear, catching the sunset, watching the kingfishers and macaws under a rare clear sky.

Once we actually started hacking the undergrowth away to make our campsite, I too started to doubt myself. The dominant plant was a palm with vicious black spines on the underside of each leaf and frond. Once hacked, the little finger-length thorns were everywhere, poking through shoes and the underside of tents. Graham popped his sleeping mattress, so he borrowed Aldo's. And popped that one too! In my head, the downriver spot started to become a perfect Shangri-La, a swaying palm trees paradise where life is sweet. Here, though, the sun doesn't set till seven, the forest is pitch-black by five. We sat in the gloom, eating our rehydrated slop meals and counting the minutes until we could head to bed.

The following morning, we were woken by rain thundering on the roof of the tents. I took the executive decision to have a lie-in. Two hours later, my desire for coffee was even stronger than my desire to stay dry, so I got up and wandered out into the wet. Rain puts a whole different complexion on a jungle expedition. You get a special kind of wet that no modern fabrics or technology can do anything about. Even the clothing you pack into dry bags in the hope that it can be your morale-boosting dry stuff somehow still gets soggy by osmosis or something.*

---

* Expedition success is all about organization, about really knowing where your towel is (to quote *The Hitchhiker's Guide*). It's about the minutiae of being able to pack exactly what you need, no more and no less. It's about being able to put everything in the exact right place, so it's to hand when needed. When it's going right, you feel like the master of your own little world, utterly in control. When it's going wrong, every minute of every day is unbearable effort. Tiny mistakes can have big consequences. Perhaps the best example of this is your 'wets' and 'drys'. In the rainforest, you have one set of clothes that you wear in the daytime, that will be sodden wet, stained with mud and blood, and they will never be dry through your time in the forest. Nothing dries here – the air is too humid, and direct views of the sun too fleeting. Your dries go on at night when you are certain you won't have to step outside again. They are your morale-booster at the end of the day, and are as

We sat drinking coffee under a leaking piece of tarp, watching all our gear get soaked. Our only hope was that we would find a campsite that had direct sunlight to dry everything.

In the canopy above us we caught the occasional bright yellow flash of the tree locals called green heartwood. This time of year they drop all of their leaves, then spontaneously erupt into a covering of yellow flowers. There were giant palms the height of an oak tree, white paper barks with scraggly open branches, delicate pink and purple orchids, thin-leafed epiphytic orchids not yet in bloom. There were trees with bizarre wavy buttress roots, like strands of warping streaky bacon, and others with lamp post-straight, thin trunks, zipping their canopy up towards the sunlight without branch or leaf. There were trees encased in a shrubbery of creeping plants, wrapped as if in green Christmas tinsel with curtains of blossom. Some had leaves the size of dustbin lids, others clusters of tiny fingernail-sized ones, but it was the vines that were perhaps the most impressive. The word 'vine' conjures an image of Tarzan swinging through the African forests (or George – 'Watch out for that tree!'), but some here are thick as tree trunks, and potentially older

---

sacrosanct as a priest's robes. On the first evening out in the forest on this trip, after we'd finished our night walk, I'd washed and got into my dries. And then someone called out that they had spotted a caiman in the river. Rather than change back into my wets, I went out as I was, and got a filthy soaking in an abortive and failed attempt to catch the little croc. I didn't have anything dry to wear for the rest of the trip.

than the trees. They spiral and twist like great ringlets from a star-let's locks, or vast boa constrictors winding their way skyward.

It's not difficult to see how animist religions have sprung into existence in all hunter-gatherer cultures around the world, from Papua to Borneo to Congo to here. It's inevitable. You rely abso-lutely on the forest for everything, and all that you see changes throughout the year – growing, giving, dying, taking. Everything here does have an energy, it has a life and a tangible existence. It's only a tiny step to see that energy as being a spirit, a ghost or a god. I could imagine there being a god residing within some of these giant ancient trees, or a nervous forest sprite inside one of the flit-ting, colourful birds.

The fish, though, would be less god, more demon. Perhaps the most common fish we found as we made our way downriver was the red eye or black piranha. These get up to three kilograms in weight – the same as a three-litre bottle of soda. Their interlocking teeth take a clean bite much as we would take a chunk out of a crisp Granny Smith. Bites to humans don't happen that often, and are usually associated with there being blood and a commotion in the water, or with people cleaning their fish in the river. In all my years in South America, I've only seen that churning whitewater piranha feeding frenzy once. It was in Venezuela in an extreme dry season. Unusually high densities of fish and low densities of prey were combined with the animals being stuck in small pools that were swiftly drying up. We dangled in a few pieces of chicken, and they were shredded in seconds.

As they stood fishing at the riverside, I asked Michel and Keiran if they'd ever been bitten.

'Just once,' said Michel. 'I cut myself shaving and dripped some blood into the water.'

'I've been bitten too,' said Keiran, 'but not by a live one.'

'You what?' I queried.

'I was cleaning a big black piranha, and its nerves kind of fired off and it took a chunk out of my hand.'

'I have a scar too!' I offered proudly. 'One took off the tip of my finger.' I showed Keiran the scar.

'It's not very impressive,' he said, turning up his nose.

'What, are you serious?' I responded. 'That's rich from someone who got bitten by a dead fish.'

The cetopsis, or giant candiru, is even creepier than the piranhas, with smooth streamlined bodies that enable them to penetrate wounds. They feed at dusk and dawn, and make a carcass look as if it's been bored through with massive drill bits. There were also darters, bigmouth catfish, characins, headstanders and tetras, as well as the bristlenose catfish, which has forked tentacles on its snout, and looks as though a whole bunch of leeches are bleeding its nose. These are cavity nesters, with males protecting a territory in a hollow in a sunken log, and caring for eggs and then fry, defending them aggressively against others.

So far on our ghost river, so magical. But then one day I woke up on the wrong side of my tent, and from there on every little thing that could go wrong, did. As I was chopping bush chillies to liven up my dehydrated mush meal, I absentmindedly sliced a chunk out of my thumb. I carried on slicing, getting blood in my dinner, and fresh chilli juice into the cut. I started sucking my thumb, which set my mouth on fire. As I was getting dressed at the riverside, I got my one single pair of precious dry undies over one foot, but not quite on the other, and found myself hopping around on one leg. With the under-crackers stretched to full elasticity, I then managed to catapult them off. They sailed gloriously over the river, and slowly sank to the bottom.

That night, after a day of constant bad calls and misery, I put up my tent five different times before settling on a spot – a spot that was once more beset by ants. This time it was leafcutter ants, traips-ing through camp in long processions. The soldiers have massive heads and jaws, and can give a nasty bite. So after all my failed attempts to put up my tent, I ended up starting all over again, put-ting up a hammock so I could be off the ground, and away from their nibbling attentions.

While our river could not be found on any map, the larger river below us was. It was called the Zand Creek, and we had pretty decent information on it programmed into our GPS, and an idea of where we would meet the bigger river in a confluence. It was likely that we'd make the confluence by the following day, but one thing was confusing us: the altimeter showed we had well over a hundred vertical metres to drop before we made the level of the Zand Creek, which was more than we'd dropped over the entire expedition so far. We had an awful lot of height to lose, and no apparent place it could happen.

Meanwhile, the way on continued to frustrate. We were no longer even trying to run the rapids, but were just getting out and wading chest-height through the stream, just to avoid the drudgery of hopping in and out of the boat every few minutes.

Camp was made early that night, on a rocky island with sparse trees. There were plenty of flattish spaces for tents, and our camp was set within half an hour. With time on my hands, I took to the tent to write my diary, and found myself looking at photos of Helen and Logan. It was agonizing. I hadn't been able to call home or even send a message. I felt so far away, so lonely, so utterly desperate to be there with them.

A little later, I spied a stunning caiman sitting upright in the river in a perfect pose, mouth open, facing upriver, waiting for the flow to drive fish into its mouth. I called the camera team to get the shot, but he dashed off at the last second. I snapped at them. Later on, wandering through the forest on my own searching for snakes, I started to regret it. I'm normally the smiliest and most relaxed person on the team. Now I was just team grump. Giving up on the fruitless snake hunt, I headed down to the river, stripped naked and waded in to lie in the fast-flowing water. In camp, Keiran had lit a small fire using as tinder a lump of resin he'd found oozing from a tree. The incense smoke wafting over the water was just like the frankincense we'd smelt in Oman just weeks earlier. So much done and seen in such a short time. Staring up at the clear night sky, I resolved that tomorrow would be better. Who else gets to have nights like these? Swinging my head torch to the right, I saw the red eyeshine from another caiman standing in the rapids just metres away.

'Happy fishing, old bean,' I said, before carefully folding my hands over my groin, and leaning back to stare up at the stars.

Perhaps the most remarkable thing about the rainforest is the sounds. First thing in the morning, you're woken by the roar of far-off dragons reverberating for miles over the treetops, and red and black howler monkeys sounding off in a dramatic, even unsettling rumble. Then the sound of the screaming piha. This is a small, drab, grey-brown bird, and you could spend months here and not see one. Like many birds here, the name is onomatopoeic. The piha says its own moniker over and over again: 'pi piha, pi piha'. Whenever I hear it on film, the sound instantly transports me to the South American jungle.* Though I love it initially, after a while it's so incessant I have to zone out.

Later in the day the cicadas start up, with their stridulating that can be louder than an industrial bandsaw; the toucans join in with their boring two-note tune, and then as dusk falls (and especially if it's rained), the frog chorus begins. In British English, we teach our children that a frog says 'ribbit'. Here in the Suriname jungle you have the insect-like ticks of the dart frogs, the strumming double bass of the gladiator treefrog, the clicking stopwatch of the cane toad and the astoundingly sheeplike bleats of the sheepfrog. It's a

---

* I have also heard it used erroneously as jungle sounds in movies set in Borneo and New Guinea!

kaleidoscopic cacophony that tells of tantalizing nights spent searching for frogs in all their glorious guises – how I spent many of the happiest nights of my life.

The next morning we all woke and prepared our boats in silence. The conundrum of how much height we still had to drop was playing on our minds, indeed increasingly so because we really didn't have much distance left to cover. The river was by now big, often 50 or 60 metres wide, deep, brown and powerful. Our nightmare scenario was that we would hit several kilometres of impassable rapids, and face dragging our boats and a tonne of equipment through the forest. That would be unthinkable.

A couple of hours into the morning and suddenly, when we took our paddles from the water, we could hear a rumble like thunder in the distance. It was more ominous than we had heard from any other rapid.

'Guess this is it,' I shouted to Michel.

'Let's get out and go have a look,' he called back.

We pulled up to the right bank, tied up our boats and trudged through the vegetation to get a look at the rapids. What we saw, I will never forget. Climbing out through an overhanging bough, we stepped onto a rocky platform. And there before us, the whole river dropped not into a rapid, but a waterfall! A falls, hemmed in by dense forest, that raged over a sequence of steps, channelled into a thundering white melee, with rainbows twinkling in the jewels of bouncing spray. The faces of my friends were almost as good as the falls themselves. Ile, Keiran, Michel and Aldo stared into the water as at an unexpected and unexplained present, looking at each other, not wanting to be the first to put into words what we were seeing. I think it was Michel who spoke first.

'This waterfall is not mapped! Steve, it's new!'

'What a place!' I marvelled. 'I don't believe it, this is completely new, no one's ever seen it before.'

'You'd think you'd have seen it from flying over,' Aldo said.

'I guess it's so hemmed in by forest,' I mused, 'it might just look like a massive rapid.'

But the more we checked, the clearer it became that this was something unique. The falls continued on beyond what we could see. Using the altimeter on my watch, I calculated a 98-metre drop from the top to the bottom. Ile was close to tears, so moved by the experience of discovering something of such beauty and majesty in his own country.

'So, Ile, what do you think we should call it?' I asked.

'Gang gan jin di wan,' he said. 'The biggest one!'

As we clambered down the rocks alongside the falls, we saw new and unfolding vistas. In some places the falls braided far into the forest. We found giant otter spraint everywhere, filled with pieces of crab shell. A pool lower down, no deeper than a hand basin, held two electric eels that seemed as long as my leg. There was one prom-ontory rock that stood out over the falls, though, that offered the finest single view of the entire place. As we walked out onto the end of it, we were hit with the powerful pungent smell of big animal.

And then, there at my feet was a big pile of poo. Fresh jaguar poo, from a huge male. It was the only jaguar scat we would see here, and it was in the most stunning viewpoint in the whole forest. Did the cat come out here to bask in the sun, cool in the damp breeze being carried by the falls, or, like us, did he just come here to marvel?

The team was bubbling with excitement as we made camp that evening. This was the kind of thing that happened to old-fashioned sepia-toned explorers with pith helmets and shooting sticks; wandering out of the forest onto a huge unknown and unnamed spectacular waterfall.

Moments like this not only make me tremendously proud, but give me hope for the future. If there remain places on the planet that are this unknown, even now, then there are also wild corners we have not yet spoilt. As more of our planet is ravaged by human hands, places like this become more precious. This jungle is a unique treasure, a laboratory for evolution, a snapshot of what the world's rainforests were once like before we started rampaging through them, taking whatever we wanted. That sense of deep privilege, of getting to stand there and touch this place with my fingertips, to immerse myself in virgin territory . . . it's something few humans of my generation will ever be able to experience.

It took us a day to carry our boats and equipment down and around the falls, then two more days of paddling to make it back to the airstrip, days and nights filled with wildlife and whitewater. But every second was blessed with the knowledge of what we'd seen, and what we might have left to find here.

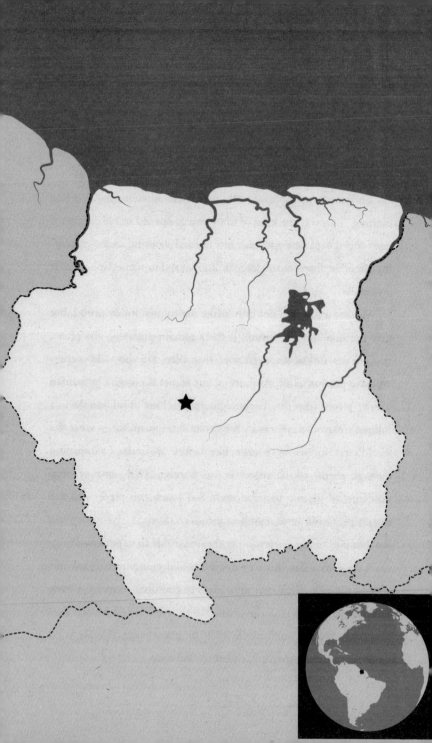

7. FORGOTTEN FOREST

SURINAME

'Rock!' The shout goes out. Through the giant rainbow a boulder swoops. We hold our breath. Just the thunder of the waterfall. And then, half a dozen heartbeats later, a boom resonates back up to us. The echo ricochets round the stadium of rock, calamitous cliffs lumbering over the deluge. I'm hanging on a line over the biggest waterfall in the nation. The ancient rock that surrounds me is too old to bear fossils as it was formed before life on earth. The falling boulder has just smashed into my ropes, lying on the rock far below me. The same ropes that we are relying on to ascend, and also to continue on down the gorge. There will be nervous moments later as I check every inch of them for frays or tears. For now, though, there are more pressing concerns. I'm hanging under the tumult, the fierce flow filling my lungs, blasting me backwards and forwards like a fly caught in a storm-blown web. Someone shouts down at me from above, but above the roar I can't hear what they're saying.

After three weeks in some of the most remote and impenetrable forest remaining on planet earth, I was dying to get to something approaching civilization and have chance to speak to Helen, possibly even to hear a gurgle or two from my little boy. But when we finally made it back to something approaching a town, we only had a few hours free, the internet connection for my one call home in five weeks was poor and the pressure was just too much. We ended up arguing. It was a bleak start to the next few weeks in the interior. There were hot, sleepless nights ahead, missing home desperately, pondering how an unrepeatable chapter of my baby's childhood was slipping away while I was getting eaten by bugs in a dark patch of forest.

Suriname had renewed my hope for the future of exploration and the potential future of tropical rainforests, but because of that everything that threatened its perfection was like a dagger to the heart. As I write this, energy companies prowl offshore, coveting the 13 billion barrels of oil located beneath the nation. At one point in our expedition, we stopped off at an illegal gold mine, expecting to see minor riverside camps as I'd seen across the border in Guyana. Instead, this was a monster operation, covering several square miles. The forest had been felled and was now sandy slop, with huge pits filled with murky orange water. The soil had been blasted away with big hoses, and then run over filters to remove the heavier flakes of rock.

The really destructive element to the procedure is that mercury and cyanide are used to separate the gold from the silt. These then leach into the environment, poisoning everything downstream, destroying waterways. On the banks of Brokopondo, the giant lake where we'd started back at the beginning of our Suriname mission, children were dying of heavy metal poisoning, as the fish they ate were full of toxins.

The gold mine itself was proper Wild West frontier territory, the hardmen who worked there in the ferocious sun covered in homemade tattoos and brandings. They lived out in the middle of nowhere for months at a time, getting paid a percentage of what they got out of the ground. Some months they earned nothing at all. And yet they also told tales of the month when they pulled up a nugget the size of a human fist, and everyone went home with $10,000 apiece. I knew I had to bring up the subject of mercury and cyanide, but wanted to tread very carefully. Turned out, I needn't have bothered. The huge foreman – standing in tatty jogging bottoms and wearing thick gold bracelets round his wrists – brought it up himself within minutes of us being there. Seems they see it as an occupational hazard.

This time our journey began in the Central Suriname Nature Reserve. The central region of the nation is much more mountainous than I had expected. Until we flew over, I'd only really studied satellite imagery and maps, and they hadn't given me the impression of a massively contoured landscape. Once in the air, though, it was as if we were soaring over a shallow green swamp, with scores of slumbering dragons lying immersed in the mire, their armoured spines soaring up above sea level. Some of the peaks breaking out of the trees were granite, with exposed rock cliff faces. My initial idea had been to try to find one of these we could climb. The main drawback was that even to get to the base of any of these peaks would have taken weeks of jungle hacking. Which was a shame. Especially as one of them, named the Devil's Egg, was a perfect sugar loaf peak, with an even more perfect egg of rock balanced precariously on top. It would have been an unbeatable(!) expedition, but it would have taken a month we did not have.

The only peak that it seemed we could visit, and would still offer meaningful opportunities for exploration, was the Tafelberg – the only tepui in Suriname. It has been visited by biologists and at least partly surveyed, as it offers completely unique flora and fauna. There is an airstrip a few miles from the peak, some rundown ramshackle shacks nearby, plus a small cleared landing zone for helis on top, but other than that, no evidence of humans for hundreds of miles in every direction.

A tepui is a tabletop mountain, made of bullet-hard sandstone, formed from the remnants of the Guiana Shield – a layer of rock that used to cover this entire part of the continent but was worn down to stumps over deep time. All that remained now were sheer-sided towering peaks with flat plateaus on the summits. Our mission would be to traverse the mountaintop and descend one of the waterfalls that plunged over the edge in search of unique plants and animals living in what we believed to be the eternally drenched world at the bottom.

Joining us on the top of table mountain were two Amerindian trackers, Mani and Uwawa, and a Surinamese entomologist named Vanessa. She was great fun to be around, super knowledgeable about the fauna of these tabletop mountains, but would clearly have preferred just to wander off and indulge her passion for water beetles than sit around watching Aldo and me coil and uncoil ropes. One thing that set Vanessa apart from other field biologists I've worked with were her snake gaitors. Whereas some naturalists working in particularly snakey environments might wear a protective gaiter of canvas, or at a pinch, leather, Vanessa had snake gaiters made of high-impact plastic. It looked as if they would repel a rifle bullet. She looked like RoboCop. I asked her why she wore them.

'You'd wear them too if you knew what was up here.'

'Really?' I asked, excited. 'You see lots of snakes up here?'

'Well, I've only ever seen just the one.'

'What was that?'

'*Bothriopsis bilineata*. I'm not sure what you call them in English.'

Eastern forest pit viper. I was surprised. 'But that's an arboreal viper, isn't it?' I left unasked the question as to why you'd need to protect your ankles against a snake that lives up a tree.

'And you'd certainly want these,' she motioned to the thick green plastic encasing her whole lower leg, 'if you got bitten by a bushmaster.'

Now she had my attention! The bushmaster is the biggest pit

viper in the world, a near mythical snake and one that every her-
petologist would kill to see. Basically, it is to snake lovers what a
bird-of-paradise is to twitchers. 'Have you ever seen one?'

'No, I never have, but there are definitely some out there.'

It seemed the mythical bushmaster remained just that. But it
still left me bemused. People give me grief for wandering around
barefoot in the jungle, and I understand that, but the truth is I'd go
insane if I had to wear boots all the time, so for me the odd splinter
is a trade-off I'm comfortable with. But every day in this hot,
humid environment encased in solid plastic? That must be hell.*

As the helicopter rounded the walls of the Tafelberg, it revealed a
land lost in time. Rock walls perhaps 100 metres high soared to an

---

* The barefoot thing is a subject I'll struggle to justify if I do ever end up getting bitten, but having spent my whole life barefoot since I was a child, I'm
never entirely comfortable if I've got shoes on. And I generally find I operate
so much better without shoes. Walking over riverine rocks, you can feel which
ones are slippery and which grip. Your method of walking is important;
treading jungle paths, you watch where you step, and you place your foot
carefully but completely. If you watch someone walking barefoot who isn't
used to it, they will hop around on their tiptoes, therefore putting far more
force down onto the balls of their feet. Even the tiniest of splinters is going to
get driven in full force. Nowadays I'm one of the heaviest members of the
crew but I can walk quietly when tracking wildlife, feeling with my soles for
brittle twigs, while others stomp about noisily in their hefty boots. It's liberat-
ing. Plus, I feel a bit like a middle-aged Huckleberry Finn. The barefoot thing
doesn't work for everyone. Aldo has been jealous of me abandoning boots for
ages. He tried it one morning in Suriname and instantly stood in a fire ant
nest. It took him a fortnight trip to stop itching.

apparently flat summit several kilometres long. I knew, though, that the illusion of flat would become a reality of gorges, creeks and ravines, and would be tough going. Worse than that, where many high tepui tops are sparsely vegetated, and the very high ones are usually blank rock with a few carnivorous plants, Tafelberg summit was fully forested. The trees were rarely bigger than an average silver birch back home, but below them was thick undergrowth. Moving around here would be challenging.

Then there was the issue of water. As we traversed the rock walls, we saw several scars of bare, well-worn rock that clearly were waterfalls for most of the year. Now, however, they were totally dry. And we were in search of the unique plants and animals that relied on water from these falls Fate, though, was on our side.

The heli touched down on top of the wall of a vertiginous gorge, which dropped down into darkness. The first thing we saw as we stepped out was a waterfall tumbling down well over 100 metres into the gorge below. We hurried down through the forest until we reached the rock bed where the river flowed down to the falls. Around it were thin, straggly bamboo and trees with white bark etched with bright orange algae and shagpile green moss, as well as one tree with waxy leaves in the shape of hearts, each the size of a dinner plate yet all hanging down vertically – almost as if an array of shields. Scurrying ticks battled up our boots to search for a place to sink their biting mouths into our flesh. The crisp drop-off was as perfect as you'll ever see: a rock ledge that just disappeared, with its brown tannic watery cargo. Over the drop-off was the giant bough of an ancient tree, its roots tangled into the rock, liana vines drooping off it in loops. Cola-brown waters flowed over a series of flat rock steps in small pretty falls, until they came to the lip of the gorge, and then they tumbled away into nothingness.

We went down on our bellies to look over the edge. Aldo made a noise as if he were vomiting, to express the sensation of vertigo. A month earlier you would not have been able to lie here as it would have been awash with water pouring over the falls. Right now, it was so flat and dry that I pitched my tent just a metre from the edge. It seemed like a great idea until I lay awake that night, listening to the thunder of water dashing onto the rocks far below, and imagined all kinds of improbable scenarios where I stumbled out of bed sleepwalking and went straight over the drop. Or more probable scenarios where a flash flood swept me and all my gear off into the void, never to be seen or heard from again.

This falls is the tallest in Suriname, and so despite its remoteness, is comparatively well known. For that reason, we had decided to find ourselves a fall on an even more remote corner of the table-top peak to descend. We trekked across the summit towards a fabled fall we'd seen on the map that might work, but when we finally reached it, the water had dribbled to a trickle. The plunge pool below was pretty much empty.

With water levels so low, the falls near camp was really our best objective. That, combined with its utter spectacle, and the assurance that no one had ever explored the gorge beyond and below, sealed the deal. It had not initially been on our list, as we had heard that one small Dutch team had previously abseiled off the top of the falls. However, on closer inspection it turned out that they had not gone down to the pool 120 metres beneath, let alone continued on down following the flow. So it remained unexplored.

We sat as near to the edge as we dared, to watch the sun setting in the distance. More dazzling still was right in the spray of the falls – three different species of hummingbird were perched on spindly branches near where the river tumbled off into the abyss. One was tiny, dark blue, but with a lurid green throat bib. The one that was most bold was grey with a brown eye stripe, green throat bib and purple iridescent cheeks. With regularity, the glossy males would spring from their perches and fly right into the tumbling water, bobbing and ducking amongst the falls, seemingly dodging individual water drops. While they danced, the dowdy females would stay back, unobtrusively watching the antics of the daring/daft males. This was a lek, where male birds come to show off for females, displaying their beauty and their skills. The female was perching and watching, while the males strutted their stuff. Often in nature male animals will do things that are clearly not good for them in order to impress the females. They'll dance in dangerous places to prove that they are able to provide good genes and are capable of reaching breeding age while doing dumb things. And that's what our hummingbirds were doing. Flitting in and out of spray that could kill them, to show the females they were well hard. Risking their lives to get the girl.

The males of our team were also about to do something

dumb – abseiling off these falls to explore the hidden gorge beneath. However, before we took the step over the brink, we decided to spend a few days exploring the tabletop of the Tafelberg. It was easier going than I had first expected. The spindly trees were rooted in spongy mossy soil – the product of any environment that receives as much as 12 metres of rainfall per year. (In the famously rainy UK, we receive about 1.2 metres. Even in Aldo's native dreich Glasgow, they only get 1.3 metres a year.) All that rain washes away many of the nutrients, and in this particular case had resulted in a relatively sparse flora, with large patches of bare rock. There were also countless riverways that would normally be flushed with water, but right now were like great paved orange brick roads across the mountain. If you'd tried to take a straight-line route, it would have been a punishing thrash. But if you were prepared to wind your way, you could cover a surprising amount of distance.

Some of the wildlife that inhabits these environments has the tendency to evolve quickly into new and diverse forms. Countless new species have been found on the tops of peaks such as these. Right by the helipad, I found a gorgeous poison dart frog, with lurid yellow above its eyes and a crazy twisted yellow line down its back. For one glorious moment, I thought we had something new. It didn't match anything in my guidebooks. A new species of dart frog on top of a tabletop mountain! But then Vanessa chipped in.

'It's *Dendrobates tinctorius*.'

'You're kidding me . . . a tinc?' The tinc is the largest of the dart frogs, and the most commonly kept by hobbyists. I was crestfallen, and more than a little embarrassed. Me not identifying the frog as a tinc (even if it was a totally different colour to any I've seen before) was the equivalent of a judge at Crufts not being able to recognize a poodle.

'That's nothing!' said Vanessa. 'The ones in the forest down below are pure yellow. They change colour every couple of hundred metres up here.'

It was evolution in motion. The frogs that lived up here were – right now – the same species as those in the forest beneath us. But over thousands of years, they remained isolated from the other groups and only interbred amongst themselves. Any colour oddities became concentrated over time, until eventually you had different colour morphs, and then one day perhaps a new species.*

One of the most unusual bugs that we saw flitting around the spindly trees was the helicopter damselfly. It's like a giant dragonfly with huge cellophane wings and an abdomen as long as a pencil but thin as a piece of string. They fly in a beautiful and strange fashion, with the wings all seeming to be going in different directions at the same time. Naturalist William Beebe described them as 'curious translucent pin-wheels'. Helicopter damsels specialize in flying up to spider webs and plucking either the spiders or the silk-bundled prey out of them. I stood for about half an hour watching

---

* Though this definition would only take effect when the two morphs no longer interbreed to provide viable offspring. A Great Dane and a pug are still the same species despite looking so different.

one hunting – eventually it flew up to a web and plunged into it, pressing it backwards. The whole web then 'pinged' like a trampoline, firing the damsel out, with a silken shrouded insect in its grasp.

The oblique edge to the falls, and the big mature trees near the edge, made it a simple job to rig up the ropes the next morning. It took 120 metres of rope to touch the plunge pool way, way below us. The next morning, with ropes attached to the sturdiest-looking nearby trees, I approached the edge and stepped off the ledge. It was a tricky start to the abseil, with the ropes taking a 45-degree bend, therefore pulling you in close to the lip. The ropes hanging below were so heavy that it had taken a good deal of effort to lug them into position. The result was safe, but didn't feel it, and looked utterly ungainly. Every tiny slip and slide on the system put your heart into your mouth with the vertigo of extreme height.

To begin with, the descent alongside the waterfall was pure box office. I rolled slowly down the lines, freehanging in space, with the mist from the bouncing water gently spattering over my cheeks in a pleasant cooling glow. It was an utter joy. Flowers dappled the green trellis, looking like snowdrops but with a purple line round the inner rim. There were also white flowers that looked like edelweiss, and misted purple saxifrage. Then suddenly there was a change in the wind and I was drenched head to toe, like I was dangling under a chilly power shower.

Next, my feet hit slimy green. The rock was slathered in nasty sticky algae, and while most of it was good and firm, there were bits that were clearly balancing on nothing more than a prayer. And some of those were substantial. A sliver of stone scythed past my face, dislodged from above by the movement of my rope. Instinctively I ducked away from it. My foot hit the wall and it wobbled. I tried to lever myself gently out of the way, but too late; a football-sized boulder shifted and fell. The rock tumbled in slow motion down below me, dropping and dropping. Finally it crashed onto a rocky ledge halfway down the abseil with a white explosion as the sandstone vaporized. The point of impact was exactly where my ropes had coiled up as they'd fallen. I groaned inwardly, knowing that if the rock had damaged the ropes, there'd be no continuing on. We'd be able to get down to the place where the ropes were cut, but no further. The expedition would have been a total failure.

By the time my feet hit the ground on the first ledge I was drenched through and shivering. There was white grit all over the piles of ropes where my dislodged rock exploded. I ran the cords through my fingers inch by inch, and though they were gritty and had clearly taken a bit of a pounding, the cores of the ropes had not been compromised. I threw the rest of the coils off over the edge and stepped back into the spray. As this ledge was about 70 metres down, from here on in the ascent would be tricky. It seemed highly unlikely that the last team that stepped over the edge would have gone further than this. So from here on in, it was all new.

For the next 20 minutes I was under the freezing spray every second. There were whole minutes where I was swung right into

the heart of the falls, and couldn't draw breath. There were honestly moments when I began to think I should turn back, that
continuing down could result in me drowning while suspended
mid-air – a dramatic but ultimately idiotic way to go.

Eventually I dropped down into a shallow plunge pool, as tannic brown as the river that fed it. Though the waterfall above was
truly spectacular, the bowl I found myself inside was extremely
threatening and claustrophobic. With rock walls towering on all
four sides, it saw light for no more than an hour a day. Even right
now at noon, the light was fleeting. I stepped out from under the
thunder and hung my shirt in the sun to try to warm up a little.
Crouching beneath an overhanging rock, I took stock.

My cameraman Graham was already starting to prepare for his
descent. Aldo would come soon after. There was nowhere here in
this bowl where we could conceivably spend the night, not a handprint of ground that was not eternally saturated. So either we had
to go straight back up the ropes, or carry on down the gorge. The
water ran down to the right and dropped over another abrupt fall;
this one much smaller, probably only 30 metres in height. There
were huge overhangs above us, with ripples left from prehistoric
seashores on their undersides. Giant male morpho butterflies in
their neon blue shimmering livery bounced around doing a may-
pole dance intertwined with gargantuan orange, brown and white
females flapping about like maidens' hankies. Beyond that I could
see the gorge winding on through a sequence of other, smaller falls,
through a preposterously tight alleyway. (Imagine two mighty

skyscrapers that soar 300 metres high but have an alley between them no wider than a squash court. That about sums it up.)

By the time the others had made it down to join me it was two in the afternoon. The sun had long gone, and I was freezing, shivering in the constant rain in just my T-shirt. Time was now a factor; we were going to have to push on, or we could end up benighted in the maelstrom. Obviously, if it rained, the sluice gates would really open and we'd get pummelled. It would not only be the most uncomfortable night in recent memory, but extremely dangerous too.

The next drop down could only be secured by tying our ropes around one specific rock, which meant we had no compromise on our line of descent. Again, it took us right through the waterfall, gulping air whenever we could. We were now into the gorge itself, and it was a chilling place, literally and metaphorically. Above was looming rock; we had to wade in through chest-deep water and then under a dark fallen tree like a gateway, creating a veil of water we had to penetrate. Everything was coated in black moss and algae. It was like a scene from *Pan's Labyrinth*.

As we pushed on, all three of us felt a sense of thrill and fear. The thrill was of the total unknown, never sure what we might find behind the next boulder choke. The fear was just as real, though; that sense that we might get trapped, might not be able to retrace our steps or might end up having to bed down in amongst the slime and greasy sodden rocks.

Our descent was now a mix of swimming, sliding and

committing climbing. The gorge dropped sharply, with big boulder chokes, logfalls and debris making progress slow and risky. Any injury now, no matter how minor, would be abysmal, as rescue would be near impossible. At times we had to resort to spelunking through gaps in the boulders.

We started to see each section as an escape room, with its own unique set of challenges. This next room had a deep pool down the centre, an unscaleable towering wall to the right, and a technical traverse to the left. The traverse was covered with piles of guano made of beetle wing cases. We assumed it was the dung of bats, but then swarms of swifts came hammering past, squealing and screaming by our cheeks and right into our torchlights. The swifts doubtless took up their night-time roosts inside cracks in the rock. And it seemed there might be room for us to roost there too. A slight overhang left a sort of tunnel, relatively dry and passably flat. Finally, a breakthrough! The Ritz-Carlton it was not, but it beat the hell out of all our other options.

Now knowing that we had somewhere to retreat to if needs be, I decided to push on, and see how much further down the gorge I could get. With slips, slides and falls now our greatest threat, Aldo and Graham hung back in the safety of the overhang, ready to come to my rescue if anything went wrong. The first drop I went down with a scramble, but the second had an overhang and then dropped about 15 metres, again inside a waterfall, and directly into a deep pool. This made things really tricky. The equipment we were wearing to abseil was heavy and clunky, and swimming in it in turbulent water was really sketchy. It also used up the last of our rope. I did, however, manage to drop down into the deep black pool, and do a laboured breaststroke downstream, the heavy metal descent devices pulling me downwards. It felt like doggy paddling with a cartoon anvil round my neck. For the umpteenth time, I found myself considering what a thoroughly embarrassing way this would be to die, drowned by my own safety kit.

After a short swim, though, I reached another drop. Beyond, there was maybe an 80-metre swim to the end of the gorge, and beyond that it looked like it might open up into paradise. From the heli flight in, we'd seen steps of waterfalls running down into the lowland forest below, none of which would ever have been stepped upon by a human being. What a prize that would be, and it was no more than a javelin's throw ahead of me. I slid up on my belly, and looked down.

The drop was short but significant, eight or so metres. Below it was deep water. And though I could have risked a drop down into it, there was no way I'd be able to climb back up. I was all out of rope, and there would have been nowhere to tie it off anyway. We were stuffed. I stood there for what felt like an age, looking at the rock, seeing if I dared to climb it. I didn't. Maybe if I were on Dorset sandstone, and a slip would have meant a twisted ankle. But I was a long, long way from Dorset. A slip would probably have meant death. Did we have enough rope to drag some through from above to use down here? No, that would have meant not having enough to get back out again. I screamed in my head, and then actually screamed out loud. The frustration! To come this far, and then be defeated by a tiny drop like this . . .

Finally, I turned back to where Aldo and Graham were watching, and made the gesture of hand slicing across throat: no go, we were turning back. I swam back across the pool and on to the rope. Except this time it was not so easy. I was trying to get my ascending gear onto the line, whilst treading water, under a waterfall. I battled against the spray and weight dragging me down, and managed to get on to the ropes, and then started ascending.

Halfway up, I reached down for my waterproof film camera to get a last shot of the waterfall. But it wasn't clipped to my belt; it had come off in the melee at the bottom of this fall. I screamed in fury. Dropping back down the line, I retraced my steps, swam back, then took off all my climbing gear. After 20 minutes of diving to the bottom of the pool searching, I had to admit defeat. It was lost. All my footage of the descent was gone. Humbled and subdued, I returned to our little overhang Hilton.

I was blue-lipped, shivering, exhausted and thoroughly hacked off with life. We'd given it our all, and come close to something really special. And now the sleep spot that had been such a relief hours before looked just like a lumpy rocky bed, only fractionally more dry than the gorge around it. As I stepped under the low rock ceiling to greet the boys, there was a huge 'thock!' explosion right behind me, which reverberated throughout the rock alley for ages.

'Holy crap, dude!' said Aldo. 'That was a rockfall right where we were just standing!'

There were white explosion marks on the rocks exactly where we'd been just seconds before.

'The helmet wouldn't do anything with a rock from that height,' said Graham. There was a sense of real fear and foreboding in his voice. That moment when everything stops being exciting scary and becomes 'Oh God, we're going to die' scary.

The night was every bit as miserable as we'd imagined. There was no point trying to put up tents or hammocks; they'd never have worked. Instead, we just wound ourselves up into little balls, trying to avoid the nobbliest rocky bits on our bony bits, wrapped in our sleeping bags. No matter where I went, there was always a drip. I managed to find one spot that was kind of OK, only to discover the drip built up over the space of about 45 minutes before falling . . . right onto my face. Just as I'd be dropping off – splunk! – I'd be dripped into wakefulness. A venomous scolopendra centipede, a small snake in my boot, and crickets and oversized cockroaches clambering over our faces didn't exactly add to the relaxation. It was flipping miserable.

Next morning, we stretched our bruised and aching bodies and prepared for the long slog back up to camp above. Real food and a decent sleep seemed a very, very long way away. It took most of the day, and by the time we got there, we were hungry, thirsty and sluggish from lack of sleep. While this section of the expedition would probably seem quite successful later, right now it felt like it had been a disaster.

When I reached the top of the ropes, exhausted, sweating,

soaked to the skin, all I wanted was a cup of coffee. I reached in my bag for my mug, but the coffee filter was up at the main camp. I put my soaking boots back on and trogged up through the forest, got the little filter, and brought it down to my tent. I set up the stove, turned it on and pressed ignite. It was damp and wouldn't fire. I searched in my tent for a lighter, but when I found it, it was out of gas. So I went back up to the main camp and went looking for a lighter. I went through all the team gear bags, but couldn't find one. So I grabbed another stove and carried it back down to my tent. I turned it on, set it boiling and took out the coffee from my bag. But the haul up had squeezed the coffee bag and it had exploded everywhere. The grounds were soggy and unusable, so back up to camp to get new coffee. When I poured the water into the mug I mistimed it and poured it all over my fingers, turning the only unchafed bits of my fingertips into suppurating sores. I switched off the stove and ran my fingers under the waterfall. By the time I finally had it done, it had been half an hour of fiddling. To get a single hot drink. I'd have paid a hundred pounds for a latte.

When you're tired and frustrated, it seems like everything on *Expedition* takes this much effort. And when it's rainy or muddy, every single step of that process is all the more unbearable. A month into the Suriname trip, and I'd reached that stage. Every tiny task was an annoyance. The dense jungle meant I was rarely if ever managing to get even a text message out to Helen – maybe once every three or four days, if I was lucky. And those text messages were typed out on an emergency device with a little joystick, each letter taking an eternity. On the entire trip we never had enough satellite phone coverage for a call.

There haven't been many moments in my career when I've overcooked it and got properly burned out. Usually it takes me months before I start to get grumpy, but I was already on the cusp of a major meltdown. When you're wet all the time in the jungle, you turn into a compost heap, a warm decomposing mess, a substrate for nasty things to grow on. My right ear was filling up with black fungal spores, and it felt as if one side of my head was perpetually underwater. My feet had furious trench foot and were pockmarked like the surface of the moon. Sheets of dead skin were sloughing off from between my toes, which were red raw where athlete's foot fungus was having a whale of a time. My inner thighs and calves were ravaged with what looked like teenage acne but were actually chiggers, a burrowing mite that lays its eggs into your skin and itches like crazy. That, plus a little entrée of prickly heat, a side order of crotch rot and a dessert serving of chafing, and my jungle experience was starting to feel like torture. If I'd been offered a TARDIS I could pop into and be home in an instant, I'd have stepped inside without even bothering to say goodbye to the team.

And then that evening we wandered out to the fall edge and sat there, watching the sunset. It's something you don't see that often in the forest, and even less so from an elevated position like this. There were dark ridges in the foreground, fading away through layers of grey mountain ranges and crags towards a horizon that must have been 60 miles away. And it was all nothing but forest. With Vanessa translating, we asked our new Amerindian friends how long it would take them to reach the nearest mountains.

'Two sleeps,' was their response.

'Two? Seriously?! It would take me a week to get there!' Aldo said.

'And what about the nearest villages? Where are they?' I asked.

'Tepu is the closest,' Vanessa translated. 'That's where Mani and Uwawa come from, with Kwamalasamutu to the south, near the border with Brazil.'

'And how long would it take to walk to those?' A protracted conversation took place between our guides, clearly very seriously working out all the elements involved.

'Ten moons,' Vanessa told us.

'Ten days away?' I said with surprise.

No.' Mani understood enough to correct me. 'Ten moons. Ten months' walk away.'

As we all pondered the realities and wonder of standing in a place where no visible human impact was evident to the horizon in every direction, Uwawa puffed his cheeks and let rip with a rasping, booming call. Mani joined him, but clasping his hands together and blowing into them. We waited in silence, and then what sounded like an echo returned. And then another louder sound from the opposite direction, and then more. Troops of spider monkeys for miles around called back to us, as we stood there listening, rapt on our mountain top.

Despite feeling burned out, it now struck me more than ever that this was the place I'd been dreaming of since childhood. I can recall reading stories – some fictional, some real – of explorers pushing into lost worlds in Amazonia, Congo and New Guinea, finding forgotten peoples with shrunken heads round their necks, wielding blowpipes and pet jaguars.

As my fascination with natural history started to develop, I read tales of Darwin, Bates and my great hero, Alfred Russel Wallace, who all came here to these same forests and discovered stunning new species every day, making the map as they travelled. I remember my bitter disappointment as a young man when I came to feel such exploration was of a different time, and that I was instead a child of the information age. But here, standing on this mountain top, I could see a thousand lifetimes of exploration still to be done. A world not yet destroyed, somewhere worth fighting for.

Scientists have described around 1 million species of animals over the last few hundred years. Best estimates are that there exist 8–20 million. Take any one of the big trees standing below me, study it in depth, and you will find new species. Without question. The overwhelming feeling was one of hope. And hope is a powerful thing.

With the Tafelberg done, our next mission would take us on another long flight across the nation, to the other new river that Michel had located for us. He, however, needed to return to Holland, and so we were in the market for new local voices to help us out. Through a friend of a friend of a friend, we heard about Diego and Jan, two Surinamese young men who worked as fishing guides.

When we first met them, I thought we were in trouble. They were in their early twenties and clearly lived to fish. They both made their best money by catching fish in far-off rivers, alive, and shipping them to China for the aquarium trade. I was initially wary of bringing them with us. However, after a long chat, where we asked them to promise they would not trap any live fish on this trip, we took them on. This turned out to be a sensational decision.

So often, poachers, hunters and fishermen know wild places and wild animals better than anyone else. Jan and Diego loved the forest with a passion, and a depth of feeling greater than my own. They were also delightful company, super hard-working, and beside themselves with excitement at getting to explore a river that had never seen outsiders before. Within days we were friends.

From the dead centre of Suriname, our next flight took us across the jungle to the far west. When we made our first nation-spanning flight, we'd looked down on the bright yellow flower of the green heartwood trees. Now, just three weeks later, not a single yellow bloom could be seen. Instead, below us, pink and purple flowering trees stood out from the green blanket.*

---

* Perhaps certain species of tree bloom at the same time to ensure mass cross-pollination. Or it could be an example of masting, where trees fruit at the

We landed at a tiny Amerindian village of no more than a hundred people. The Amerindian people are very different to the black Afro-Caribbean Surinamese locals. They know no borders, and consider themselves members of a tribe first, and nationals of a country a very distant second. They have deep brown skin and black eyes, and few of the adult men stand taller than my shoulder. Their grandfathers would have stood no higher than my chest. They smile easily, greet you warmly, and appear relatively unfazed by the strange sunburnt aliens wandering round in jungle shirts and khaki combats.

In a tiny hut in the village with wooden plank walls and a tin roof, a Baptist preacher in a red vest, sporting an extravagant 1980s footballer permed mullet hairdo, gave his sermon over a tinny PA system. In breaks during his sermon, the preacher sang karaoke pop songs while the ten or so of the congregation sang along. I sat down amongst a few of the villagers, who smiled warmly and gave me the thumbs-up.

A child not much older than my baby Logan gave me the broadest beam imaginable, before pointing wordlessly to the aeroplane; 'You came from the sky!' he seemed to be saying. It put me in mind of my first solo travels to Borneo, eastern Indonesia and New Guinea in the early 1990s. Locals asked me if I'd been to the moon, and how long it would take to walk to my village. Most had

_____

same time, to overwhelm the animals that feed on them, and ensure that not all the fruit gets eaten and some seeds can germinate.

never seen white skin before and tugged wordlessly on my blond arm hairs as if I were an alien.

Shortly after our plane landed, the heli had landed as well. The local people thronged around the airstrip, as it was the most activity from the outside world that they'd had in months. This time, rather than a tiny Robinson helicopter, we'd been relieved to see a Bell Ranger – a bigger craft, with higher payload, and much less chance of falling apart and ditching us into the jungle. Or so we thought. One of the key rules of expedition safety is not to make rushed judgements – and especially not to throw yourselves into complicated and potentially dangerous situations. With that in mind, when the chopper pilot suggested we have a look at a nearby creek in search of a dropzone, the correct reply was: 'Thanks loads, but we don't have any of our checks in place, and we haven't called anyone at home to say where we're going. We haven't even looked at the weather!'

So, when he told us he was going to go have a look anyway, and would we like to come along for the ride, Aldo and I leapt up, grabbing the GPS and literally nothing else. I didn't even grab shoes. It was as if we had just hopped into a pal's car to go to the corner shop, rather than flying in a doorless helicopter over some of the wildest jungle on earth. Within a few minutes I was very much regretting this decision.

Some would say the pilot was daring. He seemed to be deliberately trying to graze the treetops with the skids of his whirlybird and thundered down the rivers as fast as the machine would allow. The sting in the tail, though, was the storm we'd been watching approach all day. Suddenly there was a crack of lightning ahead of us, and we could see a solid black wall of cloud approaching us. In a fraction of a second we went from close and humid to freezing icy rain and vicious wind. It hit us sideways and threw the machine around like we were in a tumble drier. The rain streamed up the cockpit glass, and hammered in through the open doors. I did a quick mental stocktake. What gear had we got? None. What ways of making contact with the outside world? None. Who knew where we were and where we were going over this vast unending jungle? No one. The pilot stared straight ahead, clasping onto the control stick.

'This is normal,' he said. 'All this is normal. This is normal,' he repeated over the intercom. I thought I saw a bead of sweat run down his forehead, but it might just have been the rain pissing in through his open door.

'Normal, this is normal,' he said again over the intercom, sounding less and less convinced every time.

The cab bounced about more and more wildly. At one stage it was thrown sideways as if it had been hit by a ten-tonne truck. Aldo and I yelled and grabbed each other's legs in terror. And then we dropped. This was it, I was about to become a statistic. Or in years to come, Bear Grylls would make a programme about the survival situation where a barefoot Backshall battled his way out of the forest after eating his Scottish comrade. Or the other way around. In the pilot's desperation to get down out of the storm winds he dropped his altitude even lower. I swear the rotors must have done some canopy pruning as we bounced from side to side.

'This is all normal!' shouted the pilot. 'Where are we going to land this thing?'

'There!' I pointed. 'A landing site!'

And there on our unnamed creek was a sandbank with limited vegetation – enough to take the skids. As we dropped down, five capybara leapt off the sandbank and into the river. The skids sank slightly, then stabilized. We all took a few minutes to breathe, gather our composure and count our considerable blessings.

'That was not normal,' said the pilot.

Aldo and I were too traumatized to say anything when Jan leapt from the cockpit, rammed his hand into a hole and came out bearing an iguana, which he was to take back to the village to present as a gift. He then proceeded to take out his fishing rod and, as Aldo and I prepared the site, cutting back the vegetation, he caught no fewer than 11 giant wolf fish, one after the other. He cheerfully loaded them into the chopper, the pilot not seeming to mind their fishy goo on his upholstery. I'm guessing he took a few back home for tea.

Twenty minutes later and we were airborne again. The most severe part of the storm had moved on, leaving behind a forest that looked as if it was on fire. Steamy smoke drifted in cloudlets amongst the canopy all the way out to the horizon, the forest breathing out with the welcome blessing of much needed rains.

Morning. The cock clock started at 1am and didn't stop till seven. Dragging ourselves groggy from our hammocks was near impossible, till the sun hit and we were pretty much parboiled. The thatched roofs of the village shacks were infused with smoke from the open fires within. Mangy dogs dodged thrown sandals as they searched for scraps, wide-eyed children picked up chickens and walked round with them under their arms as if they were their dolls.

Out on the grass of the runway we stacked our tonne of kit into piles, ready for the shifting operation to get us to our remote landing site. The heli dropped us off as high up the tributary and as close to the source of our creek as was physically possible. Any higher than this and we would have been wading through a trickle.

The air was thick with green-banded urania in great clouds. Anyone would think they were butterflies, but they're actually day flying moths, named after the Greek goddess of astronomy. Perched at the riverside were flocks of jewelled green jacamar – insectivorous birds looking like long-billed hummingbirds. It was many hours later by the time all our gear and the team had arrived, and there was zero shade, so we were all burned red raw. This time, though, we really didn't want to camp here, but were desperate to get moving this same day.

I wanted to change things up a bit with this particular river exploration, so I wasn't using one of the inflatable kayaks, instead taking an inflatable stand-up paddleboard, or SUP. These have been a major leap forward in the expedition realm in recent years. They can carry a surprising amount of gear, and for me it was perfect because it elevates you a couple of metres above the surface of the water, meaning you can see further. This makes the paddleboard the perfect tool for spotting wildlife, and also for spying our way ahead when we came to areas of rapids. The team had needed some convincing that this was a good idea. They would still be in the canoes, thinking the SUP an unnecessary risk, and I'd pushed quite hard for the change, so was feeling the pressure, really hoping it would work.

Setting up the paddleboard was a breeze, and I'd managed to strap down a generator, a gear barrel, my underwater camera set-up, personal kit, food and a few other bags onto the deck. Breathing a sigh of relief, it seemed that I had it nailed. But then the eyelet that held down the gear bungy ropes just popped off. Clearly, it had overheated and the glue had failed. Now I was pretty much left with a blow-up surfboard and no way to strap my gear to it. Figuring out how to tension all the gear down on my deck was going to be challenging.

As soon as I put the board into the water and stood on it, I hit a submerged obstacle with the fin at the rear of the board and went straight overboard. I also discovered that the kayak mini-cams wouldn't attach to it, and the barrel strapped to the deck made it look a bit like it had a huge jet engine on it. Then, a little later on, I was finally paddling along serenely, standing in grand fashion and doing a whimsical piece to camera, when I clipped a rock underwater and crashed straight down into the drink. We were going to need some fine-tuning.

As we worked, we were plagued by giant sweat bees with long tongues lapping the salt from our faces. The drone of their wings was like sensory torture, and after a while seemed to be coming from inside your own head.

The creek was narrower than YiYi, but with even more spectacular foliage. Much of the riverbank seemed as if it had been decorated for some kind of festive parade. One particular tree had half moon-shaped seed pods on long wires, which dangled over the water. There were bushes with big purple kissing lips blooms, and another with hanging red Chinese lanterns. Some trees were hung with giant orbs woven like hipster lampshades – these were actually the nests of Oropendola birds, with their remarkable fluted calls. As we paddled, we became entangled in giant spider webs across the river. Their leggy female occupants were sprawled in the centre, while the tiny males sheltered off in the corner of the web, stealing the food she caught. Purple ibis, capped heron and sun bittern took flight as we paddled past.

We didn't make much ground that first afternoon, just enough to find ourselves a camp. There were a few flat rocks down at the riverside, enough for half of us, so the others had to slum it in the jungle above. Parker put his tent away from the rest of us, deeper into the jungle, as he wanted to avoid everyone's snoring.

Jan took one look at his tent and warned, 'I don't think that's a good idea, man. What if a jaguar come in here and take you?'

Parker laughed out loud. The very idea! 'We're never gonna see a jaguar.'

'Exactly,' says Jan. 'You never do see them. One day, a jaguar came into our camp and took my dog!'

Parker laughed again. Then quietly went up into the forest and moved his tent closer to ours.

Even before we'd made camp, Jan and Diego were at the water's edge with their hooks firing out over the water. After the first three wolf fish, we had to take them aside and ask them not to catch any more. As it grew dark, though, they scooped up a whole array of tiny aquarium fish from the water's edge – commandoes, odd-looking catfish and hatchetfish. Then, abruptly, a flapping stingray swam by, the size of a welcome mat with bright orange blotches all over its back. Jan reached down into the rocky shallows and scooped it out with his net. This is one of the most potentially dangerous creatures in the rivers. There are loads of them, and every time we floated through clear shallow water we'd see a little stingray squadron darting off into the shadows. At the base of the tail is the jagged barb, which can tear a nasty wound, as well as injecting a fierce venom. As rays stick to the bottom, people sometimes step on and take horrible wounds to the lower legs. (The recommended way of avoiding this is to shuffle your feet through the sand as you walk, hopefully sending them scuttling off before they nail you!)

While the rest of us sat doing all our various jobs, and planning for the following day, Jan and Diego effortlessly built a tripod of green sticks, bound together with vines, and lit a smoking fire beneath it. They gutted and butterflied the wolf fish and placed it over the smoke, and kept the fire going all night. While we tucked into our cereal bars and dried food on the river, they were chewing smoky fish jerky, looking at us as if we were heathens. Jan did try one of our dehy meals once. He nearly spewed.

We all struggled to click back into the routines that had become second nature on YiYi. It was a couple of days before I'd figure out how to get the SUP board well balanced with all the gear, while still having my binoculars, video camera, water, hat, sunblock, etc. all close to hand when I needed them. The board, though, was starting to pay dividends in the ways I'd intended. It was a pleasure to paddle, more manoeuvrable than the lumpen canoes, and I felt imperious standing high above the flow. And it really was the best tool for scouting rapids and spotting wildlife. I took to the front for the remainder of the expedition, and began to feel the SUP was not the worst idea in human history.

The creek was, in its own way, even more brutal than YiYi. It felt like little more than a stream at times, shallow and filled with obstacles, big and small. Sometimes we had to hop out to drag the boats over sand bars. Giant tree trunks across the river were too big to even consider cutting through, so we had no option but to carry the boats and everything in them over the top. Dragging the inflatables through low-hanging branches and spiky vines, we were always battling the possibility of puncturing the craft, which would have made life thoroughly miserable. The flip side of that was several times having to hack through nasty tangles in deeper water, wielding my machete above my head while treading water, dreading to think what beasties beneath might have been targeting my thrashing tootsies.

Machete-ing my way through a tangle of vegetation blocking one constricted part of the river, I must have smashed my blade into an ant nest. This only manifested itself when my SUP slid through on the other side, and the pain started. These were some of the nastiest biting and stinging ants, and I was covered in them. With a scream of anger and agony, I leapt into the water and ripped all my clothes off. Grabbing handfuls of sand from the riverbed, I started scrubbing my skin and hair to try to get rid of them as they savaged every centimetre of my body.

'This is no time for a bath, dude,' said Aldo as he cruised past.

After the smorgasbord of wildlife we'd encountered, the full daytime scorch again sent everything huddling for the shade. However, as the day wore on we saw some majestic sights: brown capuchin monkeys shaking branches at us in fury; red howlers, for now silent, but gazing down at us from the shade up high. The big moment, though, was when a giant shadow swooped over our heads, from a bird of prey whose body in flight appeared way too big for its wings. It thundered into a tree at the riverside and turned round to eye us. It was a harpy eagle. The most powerful eagle on earth, they have talons the size of grizzly bear claws, and will snatch monkeys and sloths from the treetops. This one sat openly on a branch, gazing down at us without blinking, the best view I've ever had of one.

No Name creek was proving a tough nut to crack. More often than not, we'd paddle for ten minutes before hitting an obstacle and having to carry or drag our boats. Energy was low, tempers were getting frayed. I had giant bags under my eyes and my wrinkles were starting to look like they'd been carved out with a shovel. Sunblock and sweat trickled into my eyes, which were a nice shade of fire hydrant red. My wedding ring was clacking on my carbon-fibre paddle with every single stroke I took, clickety clack, clickety clack. Normally I'd have taken it off, but I kept it there, just to remind myself with each and every paddle stroke why I was there, that every stroke was taking me closer to home.

Every afternoon we pushed on as long as we dared, always hoping to find that dream campsite of rocks in the river or white sand beach. It never came, and we always ended up sweating in the claustrophobic dark of the forest floor. The next morning we'd get up and paddle on, and inevitably within ten minutes we'd come across the perfect campsite that'd been just around the corner.

The beaches were pockmarked with perfect little sand craters, filled with water and sometimes wriggling tadpoles. These are the battlegrounds where gladiator tree frogs lay their eggs. The males will fight each other in these sand arenas for the right to breed, before the female lays her eggs inside, safe from the attentions of predatory river fish. When the tadpoles mature, they can scamper over the sand rim and out into the flow.

We had fleeting glimpses of giant river otters, periscoping their heads up above the surface to survey our boats, barking, chirping and gargling at us before silently disappearing. At one point, a big female tapir and her young child were surprised by our approach, and leapt out of the water on opposite sides. We could hear them calling to each other across the river, the youngster clearly not knowing what to do. Jan and Diego started whistling – not really mimicking the snorting calls of the tapir, just making short whistle peeps. We were all keen to push on, and after about ten minutes our patience was starting to wane. Then suddenly the whistles started coming back from the bushes. Out of nowhere, the young tapir just strolled out of the forest and into the water in front of us, where it seemed uncertain what to do next. Then, with leisurely strokes, it swam across the river in between our boats, its weird prehensile snout sticking up above the flow.

It was our last night on No Name before we finally found a good camp: a big flat sandbank, littered with iguana eggs from the lizard hatchlings that had clearly just made a break for freedom. It was utter bliss to scrub all the grime off our clothes and hang them up in the branches, allowing the searing sun to finally dry them off. Poobrella was erected at the bottom of the beach, and we all sat in the shallows, watching squadrons of macaws flying over. We chitter-chattered like we were on a beach holiday. Although the exploratory part of the river was nearly done, it had been hard, with little of the grand rewards we'd had on YiYi. We were all looking forward to the claustrophobic tangle of No Name giving way to the main waterway, seeing some sunshine and being able to enjoy the final week of paddling.

It was around noon of the next day when we hit the Lucie River. Though one of the biggest river systems in Suriname, there are no settlements upriver, and it's littered with boulder chokes. In the dry, like now, it's not navigable in motorboats. In the wet, it's a rampaging thunder of brown slurry, and would be suicide to attempt. For that reason, it hardly ever sees visitors and remains startlingly pristine. As we paddled out of the creek and joined the main waterway, it was instantly and dramatically different. The banks were a kilometre apart, the still waters reflecting the forest banks as if a perfect mirror.

We pulled up at the first giant boulder obstacle: huge cubes and balls of granite, worn smooth as silk by the passage of water over deep time. It looked like a giant troll had just discarded his Yahtzee dice into the river.

There was, however, a series of gouges in the rock, running in rows alongside each other. On closer inspection, these turned out to be evidence that people had been here before us. Way back, in Suriname's version of the Stone Age, subsistence hunters had moved up and down the Lucie River on their hunting trips, and stopped here to grind their stone axes and knife blades sharp. We were to see more of these signs as we progressed downstream, an eerie reminder that thousands of years ago hunter gatherers had stood in this exact spot, pondering the quantity of game and the abundance of potential food here.

On the other side of the boulders, a caiman as long as I am tall lingered in shallow water. As I edged in towards him, he slunk off into the water, as if annoyed that I should have stolen his sunbathing spot.

Now that we were on the main river we saw different kinds of crocodiles – black and spectacled caimans rather than the smaller, smooth-fronted caiman we'd been seeing on the smaller, fast-flowing streams. There were also terns, herons and white egrets in their thousands, and skimmers – a weird bird whose lower mandible is longer than its upper beak. They skim the surface of the river with the very tip of the lower beam grazing the water. When they hit a fish, they snap it shut and whisk off with their meal.

The flow remained slow, and so was progress. The paddleboard was much faster than the inflatable boats, which sometimes seemed to be going backwards. With the amount of mileage we still had before we made it to our destination, it looked as if we would be on the river for a long while yet.

One evening, as the daytime scorch cooled, we broke out into a river so big it seemed more like a lake, and as flat as a satin mirror. The sunset view was the most stunning I've ever seen in a rainforest. The only disturbance was occasional dips from swallows diving for a drink, and the splash of a feeding osprey that circled and called noisily over our heads. As colours bled from blue to orange and then scarlet, we spotted a corner that was marked by an expanse of flat rocks; the perfect spot for us to put up our tents.

Stingrays scuttled off at our approach, racing through the shallows trailing their wake over the perfect surface. In amongst the aquatic cabbagery, pacu streamed in flotillas and puckered peacock bass the size of border collies twerked their eyespotted tails at us. Jan and Diego's eyes were on stalks. Walking through these pools to find flat camp spaces, Jan casually swung his machete into the water, hit something hard and then reached down. He pulled out a peacock bass the size of my leg.

As the last of the sun disappeared, the colours were like a Turner painting, staining the sky and the water below with smudges of scarlet and shocking pink. You couldn't take your eyes off it. Every time it seemed you'd witnessed the best of it and got up from the hot rock to go fix some food, the colours evolved into something even more staggering. We were shaken from our reverie, though, by a scream from the bushes. Leaping to our feet, we scrambled over the rocks and through the pools into a thicket where Jan had put up his hammock.

'It's a massive great spider!' Jan said, pointing to the tree where his lines were tied. He wasn't lying. It was indeed a whopping hairy tarantula – Avicularia, the pink-toed tarantula, beloved of hobbyists all over the world. While I consider them beautiful, and used to have one as part of my home menagerie, there's no doubting what Jan would have done, had I not been around. Indeed he was brandishing his machete as if the spider might at any moment leap from the tree and start eating his leg. I knelt down and coaxed the spider to wander onto my hand, where it scampered, its gigantic leg hairs backlit by our torches.

'Man, I can't believe you'd do that!' said Jan, awestruck.

'What are you talking about?' I replied. 'You're the biggest jungle Joe I've ever met and you're scared of spiders!'

'How can you not get bitten?' he asked, bewildered. Diego likewise was shaking his head in wonder.

'You quite happily catch wolf fish and stingrays with your bare hands.'

'Well, I would never do that!'

'Are you sure you don't want to try?' I asked, offering him the spider. He shrieked and nearly ran. Obviously we all respected his sensitivity, and never mentioned it to him again . . .

The next day, our big broad river had turned into a braided mess, tangled around a slew of different islands. There were dead ends and rock barriers, rapids and log drops. What was once slow became painfully slow. We sent up the drone a dozen times to try to find the way, but really the only way on was to pick our way through the channels at water level. In the whole day we probably gained about two kilometres. Suddenly the week that we had left to get to the end of the Lucie felt optimistic.

I'd brought along an underwater film camera to try to record some of the aquatic life in the rivers. Over the years this had been remarkably successful, and we'd managed to get some incredible imagery of crocodiles, electric eels and a spectacular encounter with a giant anaconda, all filmed underwater, nose to nose in the stunning creepy world beneath the creek's surface. Here, though, the water had never been clear enough, so the camera had not come out of its box. On the latter stages of the Lucie River, though, the water was starting to clear, and when the visibility went beyond a metre I decided to put a few hours aside to get a look at what was lurking in the shadows.

Choosing a narrow part of the river, which got deep and dark quickly, I donned my mask and slipped into the brown. And nearly leapt straight back out again. No sooner had I hit the bottom than eight giant piranhas appeared within an arm's length of my face. Each one was the size of a dinner plate, and broad as a milk bottle, with burning red eyes, undercut jaw and those legendary teeth. They hung in a perfect squadron, tails finning, bodies motionless, as if waiting for the orders to rip every last sinew from my skeleton.

I shot up to the surface, breathing deeply to fill my lungs, and, making sure the camera was on, ducked back under. And saw nothing at all. We spent the next hour trying to get them back using bait, but rarely saw more than a shadow. I slipped into the next pool, ready to roll. Within seconds, a gigantic bulldog of a piranha was a foot away from me, darting in and out, checking me out as a potential food source. Then he was gone. Jan tossed in a line, and almost instantly caught another piranha. As he dragged it back to shore, a dozen other black piranhas zipped in after it, ruthlessly cannibalizing their fellow. And then they too vanished.

Over the next few days I spent a good 15 hours lying on the bottom of the creeks, holding my breath and my camera, focusing on bait in the water. On one occasion I lay next to a huge fish head for three hours, as scores of banded fish and redtail catfish tore it apart. The piranhas came in and did a brief fly-by every few minutes, but never took a bite. I got out of the water for a breather and when I returned, the fish head was gone, dragged off into the depths.

In those few dive days, I only saw two full-on feeding frenzies. Both were when a fish had just been caught, was bleeding and thrashing around in distress. Then 20 or 30 piranhas, ranking from about my hand in size to about a dustbin lid, came in at terrifying speed and ripped out a chunk, then retreated, letting the next fish in to do the same. Within seconds they had shredded the unfortunate fish. The noise through the water of their viciously sharp interlocking teeth clacking and slicing was utterly nightmarish.

Yet in all that time lying in their kingdom, wearing just shorts and a mask, right next to bleeding bait, white fingers and toes flashing, I was never bitten. I doubt many people have spent that much time observing piranha behaviour in the wild before, and what I learned was intriguing. The biggest black piranhas had a bite that could clearly have removed your big toe with one click. They were explosively quick, and you could see from marks on their bodies, opportunistically cannibalistic. However, the bigger the fish, the more wily and cautious they were. The giants would turn up first, take one look, realize something was amiss and never be seen again. The biggest ones we ever caught with a hook or net were two-thirds the size of those leviathans, but I would guess were less than half the weight and bulk of those daddy-ranhas. You don't get to be a big fish by being impulsive. What I saw led me to believe you could swim all day long in a piranha-infested pool, as long as you moved easily and calmly. But sport a cut, or a thrash of wounded panic, and you'd get bitten. And once that had happened, if you couldn't get out of the water quick, you'd be stripped to the bone.

While indulging in the world's oddest submarine safari, I also bumped into a couple of wolf fish. They were totally different to the piranha. They hung in shallow water close to the cover of an overhanging rock or some submerged vegetation. You could swim up to them so you were no more than a hand's breadth away and they wouldn't budge. They had more attitude than a fiesta queen at Mardi Gras. This made them ludicrously easy to catch. Jan and Diego were taking them out of the river as if plucking them from the freezer cabinet at Tesco. They ate almost nothing else.

Probably the greatest bugbear of any team on expeditions is food. Everyone misses things like decent bread, fresh fruit and vegetables, and it can become a real morale issue. However, on these expeditions, the problem was different, and for me far worse. Normally, we would bring a mix of rice, pasta and grains, with some tinned foods, supplemented with some local protein sources. Breakfast was usually porridge or cereal with powdered milk. For the first few days you'd probably get some fresh fruit, and maybe even eggs if you could keep them unbroken.

Then once you got into the hardcore sections of the trip – a summit push, for example – you switched to dehydrated food. These are real meals like pasta or rice curry, which have had carbon dioxide blasted over the top of them to dry them out. You pour hot water into the bag, and hey presto – dinner! Nowadays these meals are pretty palatable. However, we'd never use them for more than a few days, as they're deliberately salt- and calorie-inflated mush. On these expeditions, though, we didn't have a dedicated expedition leader back home who could have been sorting out menus and weights and measures for proper food. Instead, it was easier to count the amount of people, times it by the days in the field, and then just provide dehy meals. For the whole 25 weeks in the field.

I'd been doing pretty well up till now, but then suddenly halfway through the Suriname exped, my body just rejected it. I eat cleanly and healthily at home; bucketloads of fresh salad and veg. My stomach wouldn't take the dehy any more. I got to dinner time, and just the smell of them made me retch. And going without food when you're working physically hard for long periods of time is not a good idea.

One afternoon, Jan came running into camp. 'Where's Steve?' he shouted in excitement. 'We seen a snake!' I ran after him into the forest, and just a few hundred metres in saw Diego pointing. There was a bank ahead of me, thickly forested.

'Do you see that big vine?' asked Jan. 'It's running down it, black and yellow.'

I looked, but couldn't see anything, so started to scramble up the bank behind it.

'Tell me if it moves,' I shouted down. Snakes don't have external ears and can't hear airborne sounds, so I wasn't too concerned about the noise. I was much more concerned that my movement would spook it, and it would disappear.

'It's moving!' Jan said with urgency.

There! I saw it, sliding up a vine and into the canopy, where it would never be seen again. I just had time to register what it was, ID it, process that it was non-venomous and then leap into the bush in just the way a gazelle doesn't. I crashed into the undergrowth, slid down the bank on my backside, forward rolled and came to my feet with a fizzing five-foot snake in my hand to see Jan's face, looking as if he had just witnessed an elephant juggling fire with its trunk.

'It's OK, Jan, it's not venomous,' I said, as the snake struck fiercely and repeatedly towards my groin.

Quickly the snake calmed down – reptiles often do when gently but firmly handled. Unlike us mammals they have finite reserves of energy, and so don't fight once they perceive it's useless. As soon as it was docile, I had to gasp in wonder at the pure beauty of it. It was a tiger rat snake, glossy tar black with yellow stripes and belly. Its coloration was a case of convergent evolution – a process where two unrelated organisms evolve the same solution to the same problem* – with the mangrove cat snake of Asia.

Both these snakes live in the treetops in the tropics. Their bright black and yellow bands look glaringly obvious when the snake is taken out of its context. However, when sunlight shines down through twigs and leaves it creates slats of light and shade. My theory was that, in this dappled light, the snake's coloration was actually cryptic or camouflage.

Another example of convergent evolution could be found in

---

* Like when insects, birds and bats all evolved wings to fly despite not being even vaguely related. Or moths and fish both evolving eye-spots to intimidate predators.

the flattening of the neck to make the snake look bigger when threatened. Old world cobras do something similar, but they flatten the neck out widthways, whereas the tiger rat snake does it 'dorso-ventrally'.

We finally made the tiny Amerindian village of Amatopo as darkness fell on the penultimate day. Rarely had so much been expected of such a tiny place. The only village for weeks in every direction, it held around 30 people, all of them living in six or seven houses surrounded by unbroken forest. Their closest neighbours were across the border river in Guyana, but here the river flows deep and is many miles wide, spaced out by dozens of islands. It was a mighty divide that may as well have been an ocean.

The people of Amatopo greeted us with enthusiasm and ease, inviting us into a thatched-roof open building at the centre of the village. The children sat in our laps as we talked to the elders, with Jan acting as our interpreter. Although the headman or 'Capitan' and 'Basia' were both male, unusually for a hunter-gatherer community the men and women seemed to enjoy equal status. The headman and his wife – both in their seventies – walked over, hand in hand, and welcomed us together. They were openly affectionate to each other, which is also unusual, squeezing each other's hands with delight when I showed them photos of Logan and Helen, and giggling when I showed them a video of him farting loudly with a big silly grin on his face.

The young men all held rifles or fishing rods, and smiled

sheepishly at us, before holding out their hands in welcome. The younger women wore thickly painted-on eyebrows around their black eyes, hips swung out to accommodate a sitting infant. They laughed with each other behind their hands, looking at us and doubt-less telling jokes about the strange new men who'd turned up in their world.

Admittedly, our glee that we were heading home, escaping the forest and the dehydrated ration packs, meant we were all on fine form, in great moods and eager to see the positives in everything around us. But even so, this place was something remarkable. The people looked so utterly content, so happy and healthy. Children scampered between different parents and grandparents. Older chil-dren often took the babies. The river and forest provided everything they could ever need with absolute ease. Even right at the water's edge by the village, you could pretty much catch fish with your bare hands. Other game was child's play to catch. The groves run-ning into the forest were thick with banana, breadfruit, coconut and other exotica. They had the most completely sustainable and stress-free lives I'd ever witnessed.

'Is your home like this?' the Basia asked me, gesturing around at the forest.

'Not really,' I replied. 'Actually, I live on a houseboat.'

'A houseboat?' he asked.

'Yes, it's like a small house in a boat.'

'And is your boat above the water or underwater?'

Later on, we got on to the slightly more serious topics. The people knew how lucky they were, and did not covet the trappings of the outside world at all. They did, however, have real concerns about the future.

'This forest cannot last forever,' the Capitan said. 'People will come looking for wood and for gold. They'll bring their poisons and their greed, and they will take everything from us. We know this cannot last forever.'

I wanted to reassure him. To tell him that his village is the most remote village I'd ever been to, that the forest around it is the most perfect place I'd ever seen, on a scale I didn't know still existed, and neither his grandchildren, nor even their grandchildren need worry. But I knew that was not true. When the wreckers arrive, they come quick, and there is literally nothing there to stop them. When they come, it will be barbaric.

The following day, the Capitan and his son took us in a boat out to a small waterfall rushing through the forest. We clambered over the rocks, not really knowing where they were going, but vaguely aware they were taking us somewhere unusual. After a while we came to a spot where the granite boulders held vague scratch marks on them.

'I've not been here since my childhood,' said the Captain. 'These were left by our ancestors, hundreds of years old.' We nodded politely, but they didn't really look like much. We scrambled upriver for a few hundred metres, and by the time we had returned, the Capitan's son had taken a stone and carved anew into the etchings in the rock, revealing the shapes of dramatic petroglyphs on almost every rock. Leaping jaguars, tapis, monkeys and warriors. I had to cough back my surprise. If these were ancient artefacts, he'd just defaced them with a chunk of rock. But then the Capitan picked up a rock and started to carve himself. It seemed this was the way of things here. Every generation, someone would come by and strengthen the grooves, rejuvenating the symbols, keeping the lore alive. Antiquity was not an untouchable relic for them; it was a living part of their past, present and future, their connection to the nomadic peoples who'd been here in these forests for millennia.

8. THE DESCENT

BHUTAN

So this is what it's like to drown. Lungs choking, sucking for air, ripping cold numbing the fingertips and fogging the brain. I'm a prisoner in a cold grey alleyway where the sun never touches. Above me, impossible uninterested cliff skyscrapers glower, the gullet is whipped through by roaring icewater flood. The white thunder smashes the senses. I'm dragged below the surface again and again as if by some cruel river spirit, determined to carry out an ancient malevolence against men. I'm powerless. It's the terror of helplessness, of feeling your strength disappear, stolen by the freezing cold. The lack of breath and the frenzied exertion of fear. It's an end that takes forever, giving you time to think. To see your final minutes in slow motion, to write your own epitaph, see the faces of your loved ones and contemplate all the things you'll never get to do. This, I realize, is how I am going to die.

To the north, snow-dusted summits gave way to forested peaks, which continued to slope south to the border with India, near enough at sea level. The pine forests in Bhutan are filled with bears, leopards, red pandas and tigers. I gave up trying to memorize every kind of bird that's found here when I realized the species count is well over 700. It's beautiful in the extreme, but more than that, it is Bhutanese, and the people are rightfully very proud of what that means.

A small Himalayan kingdom sandwiched in between India and Chinese-occupied Tibet, Bhutan has had to battle to maintain its sovereignty and independence, and has managed that with quite extraordinary success. The way it achieved this was by completely sealing itself off from the rest of the world. There was no television here till the late 1990s and no internet until the early twenty-first century. In addition, the constitutional monarchy is remarkably enlightened and keen to maintain Bhutan's integrity as a genuine Shangri-La. The leaders have long extolled the virtues of Gross National Happiness over Gross Domestic Product, enshrining in law the right of every Bhutanese to be educated, literate, employed and land-owning. At least 60 per cent of the country must remain forested, and the environment's health is more important here than anywhere I've been. Plastic bags are illegal. You will not find another country in the world where more people wear the national dress, a sort of all-in-one kilt called the gho for men, and the kira for women.

The first time we came to Bhutan in 2010, I thought it the most perfect nation I'd ever visited. It's the only place I've ever been where perfect Himalayan views were improved upon by human buildings – glorious white stucco monasteries with red-tiled roofs and colourful prayer flags roaring in the mountain winds. Much has changed since then. And though I'd still say it's the most divine nation on the planet, there are cracks appearing in its perfection.

The population is increasing, and so is building work to accommodate them. 3G and Wi-Fi have arrived through much of the country, along with telecom masts, and gho-wearing locals buried in their smartphones are common sights. Rather than visiting the temples to spin prayer wheels and meditate each day, young people are using iPhone apps that spin a virtual prayer wheel for them in the comfort of their homes. And many of the policies that seemed most progressive to me now seem a little dictatorial as the outside world starts to intrude. The sale of cigarettes is outlawed here, but walking into bars and cafes in the capital Thimphu, and finding young people smoking in secret . . . it kind of feels like Prohibition.

Then there's tourism. Bhutan has always been determined not to become like nearby Nepal, overrun with backpackers, mountaineers and hippies. Parts of Nepal are so dominated by tourism that it defines them, and the King of Bhutan doesn't want that to happen here. To this end, Bhutan only allows a limited number of tourists into the country, all of whom have to pay $250 every day just to be there. They have to stay and eat in state-sanctioned facilities. Because of this, most foreigners you see here are wealthy retirees. It's a policy that's worked – these travellers tend to create the least impact on culture and to educate themselves best on how to interact respectfully with locals. As a former travel guide writer, I can honestly say that is not always true of backpackers. However, it still feels uncomfortable for an old Trot like me to see a whole nation – and one of the most splendid on earth – out of bounds to the kind of traveller I once was.

Another policy I struggled with, for obvious reasons, was the decree banning mountaineering here. Because of this, the highest unclimbed mountains in the world are in Bhutan, including 7,570-metre Gangkhar Puensum, the world's highest unclimbed peak. Obviously, this is the most coveted mountain left on the planet, and the gem we all wanted more than anything to attain for our project.

We started our negotiations with the royal household two years beforehand, offering to take a member of the royal family with us, and to reach a pre-ordained spot just below the summit, in order to respect and honour their beliefs. For many months it seemed we might have a chance, that we might be afforded that incredible honour. However, finally the word came from the royal household. It would set too dangerous a precedent; we would not be allowed to climb.

Though it was a crushing disappointment, there were plenty more options that would not be out of bounds. We turned our attention from the mountains to the rivers they give birth to. Many of these whitewater cataracts had been tamed by dams, which create the hydro-electric power that is Bhutan's biggest export. They're a big part of the reason why Bhutan is considered the world's only carbon-negative nation – sequestering more through their forest cover and sustainable power sources than they generate.

Dams, though, are the scourge of the paddling community, and of conservationists, as they can compromise waterways in some-times cataclysmic ways. There remained several huge river systems that were not yet dammed, so the next challenge was to find one that had not already been descended. In paddling the Drangme Chhu in 2010, my team had chalked up the last great descent in eastern Bhutan. The rivers of the west had been ticked off by pion-eering paddlers in the early 1980s. That left one last candidate.

Flowing through the Bumthang Valley in central Bhutan, the upper reaches of the Chamkhar Chhu were known, but down-stream of Bumthang town the river dropped into a gorge and, according to local legend, 'falls off the face of the planet'. Our investigations showed that other paddlers had bypassed the gorge, as they thought it would be 'suicide'. There was no worthwhile map of the Chamkhar Chhu, but satellite imagery showed they might well be right. The river dropped from 3,000 metres in alti-tude to 100 metres in the space of 100 kilometres. Such a fall would create a steep profile the equal of any of the world's wildest white-water descents. In addition, the river abruptly disappeared into a sequence of gorges, each of which had steep sides, and little or no chance of evacuation. It would seal us in, and spit us out. The river was too narrow and constricted to rely on rafts. Most of the time, we needed to be able to work solely out of our kayaks. First add-ition to the team, then, was James 'Pringle' Bebbington, a world champion kayaker who had been scratching a living filming com-mercials. He is also the nicest man in the world. The only thing that could potentially stand in the way of him becoming the world's best kayak cameraman might be his diet. He is a vegan who only eats raw fruit and veg, and not surprisingly is built like a half-starved whippet. As we had spent the last six months on the road without eating so much as a spinach leaf, we were concerned as to how he'd cope with two starving weeks in the wilds.

Our other young talent was Sal Montgomery, a professional paddler and physio who spends half her year guiding on big white-water in Chile. You couldn't help but be instantly charmed by Sal, a woman who may never have had a negative or judgemental thought about anyone. She's ripped, bouncy as Tigger and, like James, humble as hell. Her only gremlin was that she was recovering from shoulder surgery and was still officially in rehab. I weighed more than the both of them put together. If I got into difficulty, they sure wouldn't be carrying me.

Our lead kayaker was experienced Yorkshireman Darren Clarkson-King, who, with his long white hair and beard, was variously described as 'Gandalf' and 'Homeless Santa', but has probably run more Himalayan rivers than anyone alive.

We had additional paddle support from Bhutanese and Nepalese professionals – Chencho, who'd be kayaking with our core team, and raft guides Thinley, Kinley, Karma, Sonam, Ganesh, Chotak and Rum. Chencho was also slight in stature, but huge in heart. Before every tricky rapid, he'd take a pouch of betel nut from his life jacket and chew it furiously, spitting out great gobbets of scarlet spit. More than once we got to the bottom of a rapid and

I looked up in concern, thinking he'd had his teeth smashed in and was bleeding profusely.

There were just two weeks at home in the UK in between the Suriname slog and Bhutan. In that time I only managed to train on artificial whitewater twice, and found myself horribly out of shape.* I was overweight, fattened up by a diet of 1,000-calorie dehydrated meals, and zero fruit, veg or salad. My corpulent muscles were slack and slovenly, and I was still jaded, battle-weary and softened up by baby baths and sleepless nights with little Logan.

Of slightly more concern was that I'd been suffering from shoulder problems for several months, and this break was my first chance to see a specialist and get a scan done. Sitting in the office of a Harley Street surgeon, I was told rather grimly that I'd worn the cartilage down to the bone and would need to go under the knife to stand any chance of it working right. With that an impossibility for at least another six months, I had to make do with a horse needle stabbed into the bursa and cortisone to ease the pain.

In that precious two weeks, three-month-old baby Logan had a stinking cold and chest cough. Most mornings I'd be up from 2am to 5am, driving round local country roads with him in the

---

* At London's glorious Lee Valley Whitewater Centre, to whom I owe a great debt of gratitude. And though I didn't paddle much, it was better than my prep for the Baliem river descent, when I went down a Lee Valley rapid upside down, bouncing my head off the bottom. Hence the big scar in the centre of my forehead. A week later I did the exact same thing again!

back of the car, trying to use the sound of the engine to lull him to sleep. The eco house that Helen and I were building had hit problems with scandalous contractors, and had sucked my life's savings dry.

Like my new house, I felt like I was being held together with sticky tape and bubble wrap. I was haggard, sleep-deprived and feeling my age. I was not ready for the biggest whitewater descent of my life, and I knew it.

To then turn up at the airport to be greeted by world champion James, and fizzing bundle of enthusiasm and rippling abs Sal, was rather a crushing experience. This was not helped by the fact that when I picked my bag up from the concourse it left behind a big creamy puddle. On closer inspection, it turned out that the bottle of Baileys I'd purchased for a campfire treat had smashed inside, tainting everything from my Bhutanese bird books to my sleeping bag. An hour in the airport toilet trying to wash out the evidence had limited effect, and for the rest of the trip I smelt like off milk and old whisky.*

I spent the first days of the expedition taciturn and withdrawn, wondering if I should give up, for the sake of everyone's safety.

---

* The first time someone produced a bottle of Irish cream at an expedition campfire they were greeted with utter derision – from me more than anyone. But then we took a sip. Cream and booze; like pudding! It was a tiny cup of morale beyond compare. This unlikely tipple has been the luxury of choice ever since. I even got a bottle for my producer as a surprise birthday present and carried it around with me for the whole trip. When I finally gave it to her, she reminded me that she was lactose intolerant.

These were thoughts that should have kept me awake at night, but instead, for the first week, I slept ten dead hours a night, as if I'd been drugged. I seemed to be the first human who has ever gone away on an expedition to get some sleep.

In order to get the hang of our kit, and of working together as a team, we spent a couple of days paddling two easy rivers, the Pho Chhu and Mo Chhu, which were no more than a few hours from the capital city of Thimphu. My hope was to thrash out some of the demons with cold water in the face and a few mildly tricky rapids.

Standing on a rickety wood-slatted rope bridge across the river, prayer flags fluttering in the constant wind, we looked down at the water below us. Sal and Darren pointed out the major eddies and obvious obstacles. 'We'll get out and scout as soon as we get to a big rapid,' Darren said in his thick Yorkshire brogue. 'All pretty cruisy to begin with.'

And it was. I slid off the rocks and into the milky emerald flow that seemed to be lit from within by its own bubbling aquatic fire. I rolled once on each side to let the cold water blast the fog from between my ears. We then pulled out into the flow and were carried off downstream past a big rock that looked like a house-sized chunk of Toblerone had been dumped midstream.

We pulled into the eddy. We had travelled 100 metres. Clambering down over the boulders, river to the right, we got a look at the first rapid, and the first I was going to run on a real river in over a year.

I groaned quietly. It was a solid class four rapid, which is my comfort limit when I'm in paddle shape. The green waters dropped into a feature that looked like a playground slide into the standard paddlers' obstacle – a hole or stopper. This is formed when a submerged boulder forces water up, then it recirculates, forming a tumbling water feature that can catch or stop you. Hit it wrong and it'll flip your boat endlessly end over end like an over-tensioned roller blind.

In a controlled situation it's the kind of experience that you laugh about afterwards. In a wilder location, or a bigger hole, it is simply the most frightening experience in adventure sports. You feel as if you are being dragged down by a manifestly physical creature, coming to the surface for milliseconds to grab a breath before being whipped around again. Even the world's very best paddlers will at some stage in their career have pulled the ripcord and ejector-seated themselves out of their boat to escape the stopper. This brings only temporary respite, though, as you then face the obvious added terrors of going down a rapid underwater without a boat. Literally down shit creek without a paddle.

Here, though, the stopper/hole was just a warm-up. Below it was an assortment of big boulders in random positions causing the flow to bounce about, recirculate and thrash. It looked awful.

'So I reckon we pull out into the white horses,' Darren said, pointing into the melee, 'then river right to slip hellhole. Hey Diddle Diddle down the middle, then straight up the guts. Once you've boufed, it's all boogie woogie woogie and eddy out. What's your line, Big Steve?'

'Exactly what I was thinking,' I responded. 'You read my mind.'

In situations like this, I've always been in favour of just throwing myself into it. The more I hang around looking at something, the more the adrenalin starts to chip away at my psyche and the less focused I become. Today, though, the whole crew were just finding their feet and sorting out their systems. It took nearly an hour for everyone to be on the water and ready to shoot the rapid. By which time I'd ground my molars down to stumps and was shivering with the adrenalin.

'You chilly there, Steve?' Sal asked.

'Yeah, didn't put on enough layers,' I lied through my chattering teeth.

And then the signal went up, a couple of strokes and we were out into it. Sal took the lead, Darren just behind her. I followed a respectable distance behind, matching him paddle stroke for paddle stroke, hoping to exactly follow his line. We cut right round a boulder I hadn't seen; the flow pushed me out with it and I braced to right myself. Ahead of me, Sal and Darren were barely moving their paddles. I skimmed the side of the stopper, and then it was a cold white blur: the drop, a sideways wave and then a slide, and whitewater was towering over me on all sides. The freezing water engulfed me, and I had no idea where I was.

Then ahead of me, I saw Daz push right. He hit a boulder and flipped into a wave upside down. My instincts took over and I shoulder-charged the wave like I was running into a rugby tackle. I crashed through, and before I could even register what had happened,

I was out and in clear water. The surprise was so overwhelming that I caught the back of my boat on the still water of the eddy and capsized. Flipping myself back up, I crashed first into Darren's boat, then Sal's. We all drifted downstream, laughing and whooping.

All my life I've practised martial arts – I spent a year in Japan studying – and every time I fought, I was never any good until I'd had a good punch in the face. It was the one thing that worked for me. The shock and pain took the edge off the overload of adrenalin and focused my attention. That one rapid was just like a punch in the face. It was edgy, but I'd done it, and proved to myself and the rest of the team that I was here to paddle. For the next two days we swept down these stunning glorious green waters, and I felt nothing but joy. Finally, I could start to get excited about our first descent.

Before any big journey in this part of the world, it's essential that you visit the local Hindu temple or Buddhist monastery to gain a blessing. Every expedition I've been on in the Himalayas and Indian subcontinent has begun with such a ceremony, to the point where it has become a part of my own superstition; my very own rabbit's foot or lucky feather. They can be beautiful and memorable, a type of baptism at the source of a sacred river, or setting jasmine, rice and burning candles adrift on a boat of petals or banana leaves. Additionally, they are moments for contemplation, for remembering friends you have lost to the mountains and rivers, for asking forgiveness before an enterprise that might be your last. As the core of our team were Bhutanese and Nepali, and they wouldn't have considered an attempt of this magnitude without it.

In the town of Gangtey we walked through the village up to the central monastery. By chance, we'd arrived on the king's birthday, and the street (singular) was thronged with people who'd come from miles around to sell their wares, drink cardamom tea and socialize. Tradition dictates that visitors to the monastery should all be clad in Bhutanese local garb, so the whole crew had gone out and bought gho and kira, and had spent half the morning being dressed by our excited Bhutanese team members.

The road into Gangtey was clogged with countless buses and cars – the only genuine traffic jam I've ever seen in Bhutan. Amongst the red-robed monks and hundreds of Bhutanese celebrants were dozens of tourists, brandishing cameras and most definitely not dressed in gho. By mid-afternoon, the crowds dissipated, leaving us free to experience the monastery in the contemplative silence it would enjoy every other day of the year.

Like most buildings in Bhutan, the monastery was formed of brick walls clad in cracking plaster, inlaid with wooden-slatted windows painted with orange and black designs of dragons and tigers. Below the tiles, the roof featured characteristically stacked timbers that looked like extended Jenga puzzles. Above the main doorway, grimacing masks with goggling eyes, bulging tongues and straggly hair stared down at us. All around the courtyard were gold-coloured prayer wheels, most of them no bigger than a sweetie jar, though some were huge barrels taller than a person. As you progress, you're supposed to spin the cylinders clockwise with your hand. The hollow cylinders contain scrolls emblazoned with prayers. Each revolution creates the same effect as repeating those mantra out loud. The more times you spin the wheels, the more purification you receive.

We were invited into the head monk's private quarters, where we were greeted by a tall, shaven-headed man in his late thirties, his eyes smeary from unprotected Himalayan sun. While the monastery beyond was aesthetically stunning, his quarters were the picture of simplicity. The floor was carpeted in green baize like the surface of an overused snooker table. The shelves were lined with a fan, a veneer wireless and a teasmade alarm clock. The head monk had a heavy cold and had to break his mantra to unleash a disturbingly productive cough, before sprinkling us with jasmine-scented water and grains of rice, while ringing a bell. The bell and other loud noises are used in these ceremonies to disrupt and confuse evil spirits; the rice symbolizes productivity and the water is cleansing and purifying. As we left, we set the prayer wheels spinning, even the most cynical of us contemplating the expedition ahead and hoping for a safe return.

The drive east was in the Himalayan foothills and, like every Bhutanese journey, twisted and turned up and down the mountainsides in search of the high passes, some 4,000 metres above sea level. At this altitude, the highest forest peaks around us were shrouded in early afternoon cloud. When the cloud broke, it seemed the trees had all been sugar-frosted, frozen white by the icy fingers of the chill fog.

We looked out the window in sick horror at the piles of giant boulders that lined the roads, constantly tumbling down the hillsides to take out the thoroughfare. Building these roads must like the punishment of Sisyphus or painting the Forth bridge – as soon as you finish, you need to start all over again. We stared out the window for hours and hours as the moss-wrapped pines passed us by.

At the high passes were countless strings of prayer flags, and forests of single long thin flags the height of the surrounding conifers. Each one carried the words of the Buddha, prayers and incantations. The strings of flags are known as 'wind horses' and their colours represent the essential elements: green for nature or water, white for air, red for fire, yellow for the earth and blue for the heavens. Buddhists believe that balancing these elements brings harmony to nature and to humans who contemplate it. The flags are placed in exposed locations – ridgelines, mountain tops, passes, footbridges across the many rivers – and as the wind rifles through them, it carries off the prayers and bears them away around the world; the Buddha's wisdom taken everywhere the wind blows. Any person who is blessed by the wind will be uplifted by the prayers it carries.

These prayer flags flutter throughout the Himalayas, and are a simple, aesthetic symbol that transcends nations. One symbol that is very much Bhutanese, however, is emblazoned almost everywhere you look, and is bemusing to many foreign visitors. As we entered the Bumthang Valley, every whitewashed wall was painted with gargantuan six-foot-high penises. Of startling anatomical accuracy (apart from their scale), they sported curls of pubic hair, generous testicles and fountains from their blossoming tips. What's more, at the corner of every roof hung a carved wooden dildo, some tied up with nice bows and ribbons, others spurting fire, with wings or smiley faces. Just about every gift shop sells a bewildering array of them in every material imaginable, including candles, doorknobs and back scratchers.

Their startling ubiquity is due to the influence of Lama Drukpa Kunley, the saint known as the 'divine madman', who taught Buddhism through the unorthodox pleasures of song, dance and outlandish sexual practices. There is still a temple in the Punakha valley devoted to him, known as Chimi Lhakhang, the epicentre of Bhutan's seeming obsession with the symbol of the phallus. Legend has it that the lama used his magic thunderbolt of wisdom to subdue a demon here, and brought with him from Tibet his very own sacred phallus. This wood and silver piece became not only a fertility blessing, but an amulet against the evil eye and malicious gossip.

We were lucky enough to be welcomed into the palace where our royal contacts in Bhutan resided. The evening's entertainment involved a coterie of ladies clad in formal dress singing beautiful traditional songs and performing dances. They did this while studiously ignoring five monks, who were behaving in a very different manner indeed. Dressed in grotesque bright red wooden masks and dark green sackcloth jumpsuits tied through with rope, they were armed with giant wooden phalluses. Clearly representing mischievous sprites, they played the fool around us, thrusting their huge penises between our legs and into our faces, and mimicking lewd sex acts. It was the most unusual, sometimes uncomfortable (some of the monks were children as young as eight) performance I had ever seen. But much of it refers directly to Lama Kunley's teachings and encouragement of his disciples to embrace humour and sexuality. Pilgrims would be blessed by being struck on the head by giant wooden penises. Women who wanted children would be made fertile by its touch.

The first order of business was the most pressing. Without any decent maps, and unable to perform a helicopter recce of the river, our knowledge of it was limited. It was essential that we at least got some idea of what we were taking on, to enable us to make the right choices in how we prepared. We decided to take two days to walk, drive and fly the drone over as much of the river as we could access.

The only place to launch our boats on the river was just outside of town. Almost immediately, the river flowed into a canyon, and we would be isolated for the first day. At the end of that day's paddling it looked as though there was a crossing over the valley. That was our first stop on our recce. It turned out to be a dark green iron bridge, perhaps 200 metres above the water. The river flowed far below us, hemmed in completely on both sides by rock. It was choked with boulders and fast-flowing with almost malicious intent. To the south was a fierce-looking and sustained rapid. There was a possible way we could scramble down to the river, our first and only way to get in and out on day one.

The next stretch again could not be accessed. There was, though, a farmer's off-road track that wound its way along the hillsides to the south. It took most of the day, but eventually we found the last outpost of the district. There were three or four houses here, and as we pulled up (all with bruised butts in the back of a pickup truck) their stunned inhabitants stepped out to greet us.

The first person to greet us was Pema. His house was in the classic Bhutanese style – white walls adorned with penis figures, each corner of the roof hung with a dark wood phallus. Cows munched hay in a barn by his back door, and vibrant red chillies dried on the corrugated iron roof. His youngest son Jimmy (though I'm sure that is just our Anglicization of his name), clad in deep red robes, followed us as we headed down the mountainside. He was already a monk at eight years old, spending his life at the local monastery. We thought it best he stayed behind while we performed the tough trek, but he cried from loneliness, not wanting to be left alone.

It was a hell of a scramble to get down with no trail, but took less than an hour, which gave us hope. If the worst came to the worst, and we had to carry a casualty out, it would be punishing, but not impossible. The river, though, lived up to all our fears. We could see from a position of elevation that it was tight, constricted and full of rock obstructions. And the rapids were ceaseless, with no let-up, and precious few opportunities to carry the boats around. Rafts were going to be useless. It was either commit to kayaks alone, or head home.

Back up at Pema's house, his wife and mother had prepared us food. And alcohol. He produced an old water buffalo horn, which had been hollowed into a bottle. They insisted I take a cup of the hot wheat wine, filling my bowl to the brim. They then instructed me to put my fingers into the warm wine, and flick some skywards towards the spirit, before taking a glug. It had the scent of freshly baked bread. And the kick of a mule. They then refilled me, and I drank the draught in one single scull. Only to be refilled time and again. Sal, James and Chencho had the excuse of being non-drinkers, so Pema and his family focused on the weak one. Me. By the time I got into the back of the pickup, I was hammered. It was all I could do not to redecorate the vehicle – or my fellow passengers – as we returned to Bumthang.

The next morning, it was just the kayakers that slipped away from the sanctuary of Bumthang Valley, James, Sal, Darren, Chencho and me. The river was blurry green, shallow and broad, as easy as it gets. As we approached the entrance to the gorge, we saw two beatific boys at the riverside, with red cheeks and snotty noses. We pulled up close as they waved to us so enthusiastically it seemed they would dislocate their own shoulders. Before the river could carry us away, I scrambled through my kit to find a present for them and came across a power bar. That would do! I chucked it over the babbling river to our overjoyed young onlookers, who grabbed it with glee.

No more than 20 minutes after we departed, the hills closed in to the river, and the water constricted and sped up. Immediately the experience turned into a kind of bliss I can barely describe. The slopes were umber with a carpet of fallen conifer needles from the Himalayan pine and spruce that all seemed to cluster to look down on the strange invaders thundering through their sacred valley in brightly coloured plastic boats.

The rapids were a dream: challenging and physical, but exhilarating rather than terrifying. Cormorants perched on rocks painted white with their droppings; caramel-coloured Brahminy shelducks looked round nonchalantly, then took off in comically startled disarray when they saw us. White-capped redstarts were everywhere; this stunning little bird thrives in wild mountain streamways, and bobs its whole body up and down as it flits from rock to rock.*

As the day lengthened, the customary wind started to whip up the river into our faces. The constricted river valleys are particularly vulnerable to these winds. With 7,000-metre mountains and glaciers to the north, and the scorched plains of India to the south, temperature gradients can have winds rampaging in either

---

* There are several theories as to why so many birds (wagtails, sandpipers, etc.) that live alongside rivers practise this behaviour – perhaps to mimic the motion of the water in order to disguise themselves from predators? The redstart, though, would be very difficult to conceal, with its virulent red breast and neat white cap.

direction with a ferocity that can lift you off your feet. At our riverine basecamp on our last Bhutan expedition we had our tents take off like giant kites, before being hit by hailstones the size of fists!

As the sunlight cut down into the bottom of the gully for the last time, wind-tossed bundles of pine needles were enchanted into a million golden butterflies about us. It was breathtaking. Another river came in from the east, doubling the volume of our humble stream. Now the valley took on yet another character, as the pine slopes were overwhelmed with smoothly stroked grey rock, sculpted by generations of flood into curvy artwork.

Occasionally, we passed clusters of black crescents hanging high on the cliffs – the nests of Himalayan giant honeybees. The largest of all honeybees, their precious honey is prized by local people. It's a heck of a job to harvest it, as the nests are always hung from the least approachable cliffs. The nectar is also unusual in that, due to toxins from native rhododendron flowers, at certain times of year the honey is intoxicating. Known as 'mad honey', even a small amount can give you a peyote-like high, with hallucinations and purging effects. It's apparently possible to have a fatal overdose from a couple of tablespoons! The bees are also remarkable for the way they react to predators. When a hungry bird comes close, they flick their abdomens to the sky, which creates a wave effect across the nest. It's strangely hypnotic, and they did it in style as we paddled past.

As the walls closed in and the gradient increased, the rapids became more serious. Until now, we'd been scouting at river level, and generally just ploughing through with glee. Now, though, we needed to get out at each one, and scrabble round the rocks to look at the challenges ahead of us. Every single rapid stepped it up a notch, with some that a kayaker would call 'necky', 'spicy' or 'balls out'.

We were aiming now for the girder bridge high over the river, which we'd seen on the recce and carried the only road that crossed this section of the mountains. Our hope was that we could meet up with the rest of the team here. However, putting in the rafts was a forlorn hope; the river had been tight, filled with boulders and fast-moving. It was fine for kayaks, but would be disastrous for the rafters.

The plight of the rafts, though, was the last thing on my mind. We were cold, exhausted, pumped full of adrenalin and eager to get somewhere we could make camp, so it was a punch in the guts to see the river dropping away from us at the horizon. We pulled up to scout, and I bit my lip in trepidation. There was only one line, a chute that would send you rocketing down into a churning frenzy of whitewater. Once there, a tree had half-fallen across the flow, creating the challenge known as a 'strainer', because if you mess up, you'll get flushed against it like so many tea leaves, and will be most definitely in hot water. My first thought was that I should probably get out and carry my boat, but that didn't look possible.

'What do you reckon, Darren?' I asked.

'Not much choice, really; hey diddle diddle and straight down the middle.'

'Darren, could you for once speak in actual English?'

'Mate, nothing to worry about. Seriously, its bark is worse than its bite. And this is the crux, get down this and we're sipping chai in camp in half an hour.'

I took a deep breath. Darren was right, I had done harder rapids before. But it was tight, and I was so out of shape! I got into my boat and sat right behind him as he paddled in. Following a class paddler is almost like sitting on the back of a tandem bicycle. Matching his strokes, I could get through rapids that otherwise would be out of my league. As usual Daz took one lazy stroke to my five frantic ones. Before I knew it, I was swept into the gravity-borne sloosh, my boat accelerated and my tummy hit my oesophagus like I was on a rickety roller coaster at Blackpool pleasure beach. Before I knew it, I had smashed through a looming wall of white and had ploughed into the end of the tree. I bounced off it, and into safe water. Triumph! We all came together in the eddy at the bottom and sang each other's praises.

The light was failing as the rock walls tightened and steepened around us. It looked like we were paddling in through the open gates of an ogre's fortress, Frodo slipping in through a crack in the doorway to Mordor. It was impossibly intimidating, but it was OK – we'd done the crux rapid and I'd sailed through. Nothing ahead held any terrors for me. But then at the most walled-in section of the gorge, with rock soaring 200 metres up from the river, we approached a lip where the water fell away, clouds of steam thundering up from the abyss below. I was bringing up the rear and couldn't yet see over the drop, but Darren was ahead, right near the edge. He instantly turned back to us all, making the cut-throat sign with his hand. No way down. An impassable rapid. Chencho turned back and started paddling upstream.

I did a quick check of the surroundings. There was no eddy above the rapid and no way of clambering around the vertical walls. So we'd have to battle back upstream, then carry our boats up and out. It would take hours, and we would definitely not be out till well after dark, scrambling up near-vertical slopes with our boats over our shoulders. In expedition parlance, we were about to 'have an epic'. My heart sank in my boots. We'd got so close! The bridge where we knew we could climb out was no more than a kilometre away – a perfect day wrecked.

I was just getting set to follow Chencho when I saw James climb out of his boat and up onto a big rock mid-river. He was having intense conversations with Darren and Sal, pointing, gesticulating, mimicking the water with his hands. I turned and paddled back to the lip, where I lodged myself up against a rock and looked down. Below me was a maelstrom that made me sick to my stomach.

'There's a dirty line down the left,' James was shouting.

'Looks like we could creep through,' Darren was agreeing. Ahead of me was an edge with a small falls no more than my height. The main deluge was being driven through to the right of us and twisted into a siphon. This is the most dangerous feature in kayaking: a choke made of boulders and river-swept trees that is broader at the top than the bottom. You can get sucked in, but you'd never get out. Certain death. To the left, the vertical rock walls came down to the water and the boiling flow was being driven under the surface. James was correct; we could edge down there and comfortably avoid the lethal siphon, but if we did, we then had to pass through a chute to the left of a big boulder. It was only two metres wide, and we couldn't see where it went. It was certainly a fall of significant height into God knows what, and we had no way of recceing what lay below. Beyond this nasty invisible bottleneck was a significant distance of churning cataract hemmed in by towering rock walls. To me it looked like suicide.

'If we can do a dirty left, we'd be in and past that,' Darren mused. 'And that walk-out would be a ball ache.'

'Are you serious?' I said.

'We could sneak it,' said James. 'It's probably not all that bad.'

'I could run it first and check it out?' Sal offered.

'But what happens if it's shit?' I stammered. 'Who's going to help you?'

Sal shrugged in a way that said, 'This is my day job.'

Darren looked at me and held up both his hands. 'Trust me, Steve,' he said. 'You've got this.'

I looked at Chencho. He was taking a wad of rolled-up leaf and betel nut and chewing it furiously, teeth turning dark red. He didn't look convinced at all.

Sal went first, shuffling off the rock ledge and down into a stillish pool under a small waterfall. She expertly turned her kayak with a corkscrew movement of her vertical paddle, and was out into the melee. Her tiny green boat bounced down the boil like she was trying to tiptoe through a nursery without waking the baby. Abruptly, she hit the drop and was gone. It looked as if she'd fallen through a trapdoor. Just seconds later, she bobbed into our vision downstream. She held her paddle aloft to signal she was OK. Then Darren followed. He repeated her manoeuvre exactly and then was down the chute and out.

I felt sick. I edged over the rock lip and dropped. It wasn't as neat as the others. I got caught up in the pool, lodged between rocks, and it took a bit of doing to shuffle myself out into the boil. Lacking the precision of my teammates, I was caught more by the churning flow and hit the wall, not hard, but enough to knock me off my stride. I looked right at the siphon – end of days – off to the side of me. One more stroke and I'd be safe.

But I'd forgotten the golden rule: look at where you want to go, not at what you want to avoid. Distracted, I'd not taken in the chute until I was right at the top of it and then over it. I gasped. It was a vertical drop, three times my height and only a couple of metres wide, but almost the entire power of the river was being funnelled through it. I threw my hips forward to accelerate the boat, but still felt myself pitching straight down. In a half-breath, I corrected by throwing my torso back onto the deck behind me. And then I hit the hole below the drop. My boat instantly flipped. I rolled once. It flipped me again. I gasped for air, it flipped me once more. And then my amateur-hour paddling took effect. By throwing myself backwards, I'd weighted the rear of my boat and lost all my speed. The river's powerful hands gripped the back of my kayak, and inexorably reeled me in. The hole had me, and there was no way I was getting out. I flipped over and over and over, all was white and roar. I was in trouble.

Twenty tonnes of water were flooding into this goblet of rock every second. No human being is strong enough to battle against that kind of force. Upside down and panicking, my next attempt at a roll was pathetic and didn't even give me time for a breath. My next was weaker still. I'd been under too long and was dangerously close. Reaching forward, I grabbed for the front of my cockpit, found my ripcord and pulled. Out! Free! The cold embraced me, and gripped me in a frozen squeeze.

I battled upwards, fighting for the surface, but the hole still had me and pulled me down under. I had to breathe! Down once more, then for the briefest of milliseconds my face broke the surface and I reached for air, and I was under again. I have no idea how long this pummelling lasted, but it was minutes. Then my boat smashed me in the head. I clawed for it and used its buoyancy to drag myself to the surface. A proper breath this time. And then the boat got sucked right back in under the fall, and I was under. Often a boat – with its superior buoyancy – will pop out of even a good-sized trap like this, and staying with it is a good idea. This was not one of those times. The recirculating feature had phenomenal power, and was just sucking in me and the boat, tumbling us like an infernal washing machine. I let the boat go and was under again. Next time I surfaced, I gasped for breath, and it didn't sound like me. It was more like the death wheeze of a dying man. It was the first moment that I realized I was drowning.

This cycle repeated several times, and then a surge of force slammed me out against the rock wall. I swivelled and clawed the rock. My fingertips caught a tiny lip. I pulled, and gained a metre, edging down the wall. I got a better fingertip hold. Hanging on as the water pulled me back towards the hole, I could breathe but was on borrowed time. There was no doubt I would not get out of this situation on my own.

'Help,' I gasped, and then louder, 'Help me!'

There have been several times in my life that I've come close to death, and the one thing they have in common is that they were all over in an instant. My 2010 rock climbing fall seemed to take an eternity, with my life flashing before my eyes, but in reality was over in a second – and then there was just the pain. Being charged by a bull elephant, or having a crocodile swim straight for me underwater – in retrospect they seemed long because perhaps they have had a disproportionate effect on my life, so I've stretched them in memory. In actuality, they were over in the blink of an eye. Drowning in that dark thrashing canyon took five minutes. That's five minutes where you have time to think, to realize the consequences and fully take in the horror of what's happening to you. I can remember, after shouting 'Help', thinking to myself: *I've never shouted 'Help!' before! Forty-five years old and I've never shouted 'Help'.*

Wow, that's weird. You'd think it would be something you do all the time. But then – no, maybe it's not that surprising after all.

I saw Helen's face, and little Logan. I visualized them receiving the news that I had died in some gloomy gorge in the Himalayas. In my head I played back a video Helen had sent of them doing 'Row, Row, Row Your Boat', Logan's little face lit up with joy, trying so desperately to make his first giggle. I'd never see that first baby gurgle, never get to see him grow up. All this and more. My parents, my little nephews never seeing 'Unckie Steve' ever again. Bleak scenarios playing out in detail. When every breath feels like your last, five minutes is a very long time.

My fingers were ripped from the wall, and I was back in again. Several more cycles in the washer, and I was out. I grabbed for the wall, but I was weaker than I've ever felt before. My fingertips were bloody, raw and icy cold; I could hang on for just a couple more minutes and then I'd be back in the hole again, and this time I knew I wouldn't be strong enough to get to the surface. In my brain I had time to process the fact that I was running out of fight. I couldn't get out of this myself.

'Steve!' I heard a high-pitched shout above the roar. I turned. Sal had somehow battled back up against the stream, climbed out of her kayak and scrambled to the other side near-level with me.

'Help me!' I shouted again. My boat bounced close and I reached for it.

'No!' she yelled. 'It's in the hole. Leave your boat, swim into the current!'

'I can't,' I screamed back. 'I tried.' My fingers were finished. Sal had gone to all that effort and I was going to drown right in front of her. 'Line, Sal!'

Sal dashed back to her boat, but for some reason didn't return. I learned later on that the clip she used to keep her safety line in her boat had jammed. She'd had to rip it out with her teeth. The delay can't have been long, but felt like forever. And then she was there on the rock, prepping her throw line bag. She only had one throw. If she'd missed me, I wouldn't have lasted long enough for her to prep the bag for another go. I can't imagine the pressure for her, trying to hit the tiny orange blob of my kayak helmet, thrashed in the maelstrom, right at the furthest extent of her throwing range.

She didn't hesitate. She threw the bag and it sailed upstream of me into the hole, just out of my reach. I let go of the wall and was sucked under, but my flailing hands hit rope. I grabbed and gripped, and almost instantly pendulumed out of the hole, out into the downstream flow.

The elation was short-lived. Now I was being dragged under against the pull of the rope, and wasn't strong enough to hold on. My hands lost the line, and I was off into the rapid. I have no real idea what happened next. I know I ripped down two or three rapids, pinballing off rocks, battering myself black and blue. But then Daz's blue boat was ahead of me, tucked into a precious alcove. 'Swim, swim, swim, Steve!'

I lashed out with everything I had left, but had no energy. It felt like my arms and legs weren't attached to my body. And then Daz grabbed me and hauled me out, and into the biggest and most welcome hug I've ever had.

It wasn't over by any means. It took 20 minutes for the hole to finally release my boat. Daz raced off after it, but couldn't catch it till way downstream, so I had to climb over two more rapids along a nasty rock traverse, then jump back in and be dragged by Sal in her boat through another whitewater section.

My paddles were gone, I was frozen and my fingers dead claws from cold and rock burn. But that was just sand in your shoe, grit in your eye. I was alive, and however miserable the next bit was, I was going to get home. We were still in the gorge, though. There was no climbing out yet. My crash would have to wait. Shivering so much it hurt, I got back in the cockpit, and with the help of Darren's spare paddles, the three of us continued down.

Much later, round the campfire, the mood was sombre. I was physically and emotionally raw. One slightly too long hug from one of the team brought me to unexpected tears. I called home, and Helen told me about missing her junction on the motorway and how Logan hadn't slept well that night. I didn't say anything. How could I put what had happened into a few words on the sat phone? I just told her how much I loved them both. Then I sat alone in the darkness with my head in my hands. James and Chencho arrived in camp much later. They had seen what had happened and cut their losses, backtracking and climbing out, boats abandoned in the forest.

The whole situation was given extra meaning when Sal told me the exact same thing had happened to her just a few months before. She had been caught by a near-identical hole, dragged in, held and popped out of her boat. She had lost consciousness, been dragged out seeming dead and was resuscitated by a friend. It was clearly traumatic for her to retell the tale, and even more so for her to have to watch it happening again to me.

The question now, of course, was whether to continue with the rest of the expedition. The sensible call was to abort. The river was unrunnable for rafts, which meant most of the crew could have no involvement in the expedition. The level of commitment and danger was unacceptable. It was clear we did not have the time, logistics or (for my part) the skill to do the entire river all the way to the Indian border. We should call this one a bad deal and bail. I never wanted to kayak ever again. The idea of getting back down into that gorge made me so frightened I just wanted to curl up in my tent and cry. There was also no doubt that physiologically I had taken a proper kicking. I'd separated my shoulder, had a purple bruise the size of a tea saucer on my hip, and, though I didn't feel it yet, every muscle was exhausted to the point where it would take me weeks to fully recuperate.

But there was a bit of me that didn't want to let it beat me. I knew that if I gave up now, I'd never be the same. I was pretty damaged, but if I called it off, the cut would be even deeper. Without a decision made, everyone started to head off to their tents and sleep. Nobody wanted to be the person to call the expedition over, or likewise to put any pressure on me to continue. Once everyone else was gone, I sat beside the fire with a bottle of single malt and polished off far more than I should. Not surprisingly, I woke in the middle of the night with a filthy headache and had to go out into the rain to pilfer through people's gear to find a few mouthfuls of sweet marsala tea to quench the acrid thirst.

Sleep blessed me for barely minutes. Lying in the tent with rain hammering on the fly, I prayed for morning, then when it came I prayed it would get dark again. At breakfast, the rest of the crew laughed and joked like it was just another day. Didn't they know? Didn't they realize I'd nearly died? That nothing would ever be the same for me?

Things between me and Sal were very different. She knew; she was the only one who'd really been there. I say someone is a 'life-saver' every single day, mostly when they've brought me a chocolate Hobnob to go with my cuppa. It's such a cheap cliché. But when someone saves your life, when they genuinely protect you from the reaching fingers of death, it's something very special. The five of us had all become super close really quickly, as you often do on expeditions when you're placing your life in someone else's hands, and vice versa. But with Sal, it would never be quite the same. The closest comparison was my pal Tarx who carried me out from the cliffs over his shoulders after my climbing accident, with my back broken and smashed foot gushing blood. I know that every time I see him for the rest of my days I'll get a surge of gratitude for what he did for me. But this is the first time someone has actually saved my life. The gift Sal gave me is one I'll never be able to repay.

It really was a beautiful morning. After the last night's rains, the mists seethed through the pines way above, shafts of dusty sun beaming down on us, making our tents steam. Bronchitis-bothered ravens cawed and cronked, with hordes of Himalayan chough dancing and prancing over the peaks. I'd nicknamed the Chamkhar Chhu the 'Dragon River', a slumbering monster of staggering jade green beauty, with rapids that often steamed like the tendrils of smoke from a monster's nostrils, and other times roared. We carried our boats and gear down to the riverside in solemn mood. James and Sal could see that my mind wasn't right, and were both super keen to let me know I didn't need to do it, didn't need to prove anything to anyone. That had the opposite effect, and made me feel even more pressure. It didn't help that we were starting right where we had finished the previous day, down in the dark and cold of the gorge. It also meant putting in right at the top of a class four rapid, and there was no easy way down it.

I think I barely muttered two words to anyone until we were down on the water. To one side, a delightful waterfall stream dribbled pleasantly in to join the main flow. On the opposite side of the river was the ugly scree slope we had scrabbled out of the previous day. And then, within a stone's throw, the grey cliffs towered. It felt more like a mid-winter Welsh slate quarry than a Himalayan Shangri-La.

Then suddenly Aldo shouted: 'Stevo, wolf!'

I turned to look in the direction he was pointing. On the bank opposite us were two canid forms scampering over the messy scree. They had rich russet coats, seemed to be about Alsatian-sized and had thick, brush-like black tails.

'Dhole!' I exclaimed. Himalayan wild dogs, incredibly rare – in all my years in the Indian subcontinent I'd not only never seen one, but I'd never met anyone who'd seen one. And here were two, right in front of us, just when I was in need of a little courage. The first dhole stopped for a few seconds and stared across at us, before trundling after its partner, its thick black tail bobbing behind it. It felt like an auspicious symbol. Like the nature gods telling me it would all be alright.

'Right, Darren, let's do this,' I said. 'Do you mind if I stick right on your tail?'

'You do whatever you want,' he replied. 'Just remember, this is no harder than stuff you were smashing yesterday. Just give it beans.'

I assumed what he meant from that was that I should attack the rapid. And so we did. As with literally everything on this river, it was harder than it looked. In fact, I hammered into what was swiftly becoming this river's signature move, a Mortal Kombat fatality stroke that lulls you into a false sense of security, before raging up and punching you solidly in the McNuggets. Without exception, every time Darren or Sal assured me it was going to be OK, it wasn't. It felt like I was bouncing from certain death to constant peril with every serpentine snakey turn of the flow.

My boat powered towards a midstream boulder. All my focus was driven towards avoiding it, doing everything I could not to be pinned against it. (Of course, with 50 tonnes of water a second driving that way, you either bounce or you slide around it. You're much more likely to get pinned or folded around something much smaller.) The concentration on one obstacle, though, means you fail to fully comprehend the next, which is inevitably worse. Drops came left, right and centre; contradictory swells of flow and sneaky submerged 'screw you' stones bump your boat and spoil your line. It felt like driving a banger in an old destruction derby.

World champion James came out of that rapid with his eyes on stalks, face ashen.

'That was brutal!' he said. 'This river is insane.'

Sal ripped a turn into the eddy and looked at me with concern. 'Are you OK, Steve? This is pretty full-on. You're doing so well, I'm proud of you!'

Any other time I'd have made some comment about being patronized by someone half my age and a third my trouser girth, but I was too gripped, frankly too terrified. And Sal had a free pass to say whatever the hell she wanted to me for the rest of my life.

We carried our boats once, down past a double drop of two falls that really didn't look that desperate to me. In fact, I thought it might offer me a nice spectacular waterfall to power over and look like a total hero. Perhaps here was my shot at redemption? Then Sal urged me to watch halfway down one of the falls for a minute or so. It was like one of those Magic Eye pictures, where if you stare long enough, it starts to transform into a 3D space rocket. Halfway down this particular waterfall was a submerged and hidden rocky snag. If you'd steamed over the top very slightly to the left, the front of your boat could have lodged beneath it and stuck. It would then have instantly filled up with a quarter tonne of fluid, locking you inside to a watery end.

'Good call, Sal,' I murmured, shouldering my boat, musing quite how much I had left to learn.

There was certainly no need to push uncertain goals in search of good challenges. The river was relentless and remorseless – more so than any we had encountered before. Chencho counted 17 rapids by mid-afternoon, and he was counting one rapid as being a continuous stretch of whitewater between rests. Some of those could easily have been broken down into five or six separate rapids. In this day and age, people get excited about running a single rapid no one's ever done before, and here we were, nailing dozens of them!

But then out of nowhere we hit our first stretch of flat. It was bizarre after the constant thunder of the whitewater, and the quiet was unsettling. The still of the air, the fact that for the first time we weren't screaming at each other but could talk normally. It meant that automatically we all switched to hushed tones, as if we were talking in church.

Two dippers bobbed at the riverside, dumpy brown birds like overfed wrens, set apart by their habit of hopping into the water every few seconds and immersing themselves in their hunt for aquatic insects. A serpent eagle circled overhead, gazing down on us.

We paddled in formation into a wide section with a rare rocky beach to river left. Beyond that we could see a jumble of caravan-sized boulders, and as we looked downstream we could see the tops of the trees were below us. This was a sure sign that there was a big drop ahead. We all pulled up the side and went to take a look. The first section of the rapid looked awful: powerful, tangly and with scores of potential siphons. The real concern, though, was that the rapid was long – perhaps half a kilometre – and could have turned into pretty much anything.

It would have taken us hours to recce the whole thing, and the light was already fading, so we took the unusual step of flying a drone over the rapid to get a bird's-eye view. Darren had been carrying a tiny drone in the back of his kayak for just such an eventuality – a lesson learned from the previous evening's near tragedy. What the drone showed was not pretty. There were pieces we could have paddled, but they inevitably led into sections that would without question kill you.

'We're walking, kids,' shouted Daz.

Walking doesn't really accurately describe the activities of the rest of the day. The portage involved climbing and slithering over rocks, and through tangled bushes teetering over horrid drops – and all of this done with a plastic kayak over one shoulder and a paddle in one hand. For the first time we wished we were carrying a machete, as the vegetation tore at our drysuits and vines reached out to grab helmets and handles on boats, slowing our progress to a crawl. Though less frightening than the river, it was much more physical, and made it clear my tank was empty. I'd not drunk anything all day, and had eaten nothing more than a handful of nuts. I was essentially trying to drive a 4x4 with only a half cup of pineapple juice in the fuel tank. I was also becoming aware of how much my near-death experience had taken out of me. Every muscle was fatigued, every sinew stretched. It was a horrible feeling, knowing there was an awful lot of scary ahead and I would be taking it on with a body that had had enough.

At the water's edge was a broad flattened blueberry pie, full of seeds, but otherwise like a circular purple cowpat.

'Bear!' I exclaimed. It was the scat of a Himalayan black bear, and though already dried and lacking odour, it was a sure sign that this hulking homunculus had been here – and had probably clambered along the same tenuous trail as us.

Eventually, as the last light faded from the valley and the sky above us darkened, we put our boats back into the water and took to paddling once more.

'Where you at, Steve, on a scale of one to ten?' Darren asked.

I pondered. Should I tell him I was hovering around three? 'I'm a solid six, Daz,' I said, lying through chattering teeth. He looked back at me with a face that said: 'You're a six like I'm a supermodel.'

'Sal, what about you?' he asked.

'Yeah, I'm about a six,' she pondered.

Really? I thought Sal was indestructible, and she was only a six? I was both heartened and slightly wary, knowing the river was giving everyone a bit of a kicking, and not just me. Darren nodded assent, and pulled back out into the action.

There were still three more rapids, any one of which would have ranked amongst the hardest I've done. There was no time to stop and scout them, no time for half measures – it was absolute commitment or nothing. We needed to reach the assigned end point before it was completely dark or we would be sleeping out in the bitter chill of the highland night in our wet clothes, without food or sleeping bags and tents. Such a prospect focused everyone on getting down, safe and soon.

Half an hour later, we rounded a bend to see a narrow plume of smoke rising from the riverside and the matchstick man of one of the rafters standing on a high boulder awaiting our arrival. The rest of the team had put in a Herculean effort. They'd driven in 4x4s an hour and a half down a rutted donkey truck to Pema's house. Then, enlisting the help of everyone Pema knew, plus a caravan of mules, they'd carried all the kit another hour and a half down the steep mountainside (with no trail) to get to the water's edge. They had arranged a cook tent, proper food and hot tea. I didn't even have to filter my own water. The joy at seeing everyone, though, was tinged with one major negative: they'd brought the rafts.

I'd been worried for the best part of six months that this river was not going to work with a film crew. There could be a class five rapid round any corner, and it was not a river we should be pitching anyone other than experts at. What we'd seen over the last week had made that even more obvious. Rafts provided essential support for the filming – as well as carrying tents, food and medical equipment – but they could not have got down the sections of river we'd run so far. Impossible. There had not been any stretch of even a kilometre without obstructions that rafts could not avoid. Also, I've been paddling 30 years and had come within minutes of death here. My already heavy mood turned thunderous.

As the team saw it, we only had three and a half kilometres to go until the last possible takeout. This was another remote beach, which could again be accessed from high on the mountaintops. After that, it was 60 kilometres down to the next accessible point. They reasoned that we had to at least try to run this short section, and see if it was possible. I thought we had proved that it was not, a thousand times over.

It's a tricky situation being a part of a team like this. Surrounded by world champion paddlers and people who kayak for a living, I recognize my inferior skills and tend to defer to those around me. Not surprisingly, then, my position as overall expedition leader began to look pretty tenuous. Add to that the absolute spanking I'd had on this trip, and my withdrawn and gloomy mood, and inevitably no one was really listening to what I had to say.

The thing about whitewater that is more dangerous than anything is that it takes you by surprise. Riding a big rapid on a known river is one of the most fun things you can ever do. You could do it in a raft without ever having picked up a paddle before. The exact same rapid on an unknown river can have a feature you don't see that will definitely kill you. This river was running a comfort-able 50 cumecs, or cubic metres per second. That's 50 tonnes of water passing you with every heartbeat. No one can fight that. You can run a river one day and have nothing but giggles. Or you hit a rock, bounce out of the boat and have an experience like I did that will change your life. I didn't sleep a wink that night, for seething and raging at everyone around me.

The next morning it got worse. Right outside camp was a beau-tiful wide rapid, tailing out into a glorious leaping wave train. It wasn't difficult, but had a section in the middle that was constricted, with a significant drop dead in the centre of the river. Most of the water went through that, but due to a big boulder choke just upstream, it was really tricky to make that line.

As a kayak team we all went through it, and did not exactly cover ourselves with glory. Darren, Chencho and James nailed it, but I bounced off the boulder choke, missed the line, flipped, righted myself and went down the dangerous side of it backwards and struggling, narrowly avoiding really hurting myself. Sal hit the perfect line, but at the bottom of the fall was ruthlessly flipped, dragged back into the hole, and battled and battled to get herself out. She finally emerged, blue-lipped, shaking and clearly furious with herself. In typical Sal fashion, she decided to run it again and do it better. Exactly the same thing happened, but slightly worse. It looked awful, and we were all waiting, throw bags at the ready to save her. We didn't say anything, as we knew it would have infuriated her, and she'd have just gone back and tried again.

This, though, was no big deal for the rafts. Chief paddler Kinley had said this waterfall was 'unrunnable in a raft' and everyone had 100 per cent agreed. There was a nice calm eddy at the base of the rapid, and the raft team had all agreed the night before that they'd launch there. So that was that then. No cause for concern.

Until that morning, when I watched the Bhutanese rafting boys put the rafts in *above* the fall, and start prepping themselves as if to start paddling.

'What's going on, Darren?' I asked him, from our vantage point downstream. 'They're not running it, are they?'

'No, mate,' he said, 'they're launching down here. Least that's what they said.'

But they weren't. Apparently, they had changed their minds.

'Should we say something? This is nuts, isn't it?'

'Well, put it this way,' Darren said, 'they're the best in the business, and know what they're doing. But if they wrap that raft, they'll never work again.'

My fury and confusion were momentarily calmed as the first crew pushed off. They were all clad in identical blue and red kit, and paddled to barked orders with mechanical precision. They were evidently a superb team, who'd worked together for years. It was actually incredibly impressive, like watching the Red Army doing marching drills, or the Red Arrows flying loop the loops. Poetry in motion. They hit the midstream flow, beat the first drop and then struck for the line down the fall. And they bounced off the boulder just as I had done, and pinballed all the way down exactly the same rock-strewn mess. It was a disaster. Somehow, they avoided all the obstacles and bobbled unconvincingly out into the eddy, whooping and hollering.

OK, so now the other boat would have to give it up, right?

The second crew launched closer to the main crux fall, with less chance to build up speed. They were not clad identically. They did not paddle together in unison. In fact, they looked all over the place. We watched in horror as they missed the line completely, bundled messily down the same way as the last crew, and went sideways into a sunken boulder. The entire raft was instantly consumed. Fifty tonnes a second poured into the inflatable yellow dinghy, completely immersing it.

Shore-bound paddlers lunged for throw bags and shouted commands as they scrabbled over the rocks. The rafters on board fought to get to their feet and out of the tumult, struggling to hurl precious kit back to land and battling not to be dashed on the rocks. It was chaos. Eventually, after about 20 minutes, the stranded paddlers managed to bounce the boat free. No one was hurt, no kit was destroyed, and there is an argument that if you were going to have a disaster, then there – where safe rescue was possible – was as good a spot as any. But I was livid.

It was the angriest I'd ever been on an expedition. And it was just the beginning of a very long day. Around the very first corner was a protracted rapid that we kayakers managed to creep around, but the rafts had to 'line down', lowering the boats on ropes slowly along the side. The kayaks paddled down in under a minute. It took the rafts three hours.

Around the next bend, though, was the kind of surprise that was the very reason we'd all come to Bhutan – arguably the most perfect, dramatic and delightful rapid of the whole trip. The canyon here was like a picture postcard. To river right was a cliff face in the shape of a tsunami wave. Forty metres above the water hung a slew of Himalayan honeybee nests. Below them, the green waters funnelled into a log flume, a glorious slidy merry-go-round, with a roaring furore beneath.

It was a little after midday, and the whole thing was glinting with rare sunlight, which was only penetrating down into our little chasm for an hour or two a day. We thundered down it one by one – first the rafts, yelling and screaming with pent-up excitement and relief, then Daz, Sal, Chencho and James. When my turn came, I cocked my paddle blade into the water as if trying to jam Excalibur back into the rock, and focused on pulling the whole boat up to and past the blade. The kayak leapt forward, gathering momentum. Then down the slide and I punched through the whitewater, hollering like a lunatic. It was the first time in the whole trip that I'd allowed myself to enjoy it, and to believe we might actually make it.

The rapids just kept on coming. We hit three or four in a row that I would have called a fairly solid four for us in the kayaks, but considerably more awkward for the paddlers in the rafts. Their progress was painfully slow, and as we were trying to move as a team, it meant hours and hours lying in the sunshine, trying hard not to freeze, getting bored senseless as they lowered the rafts down the rapids on ropes.

Countless times when we did get back on the water, I bounced through a cataract, seeing off obstacles to either side that could have seriously undone me, and only beating through them by pure chance. Just as on our Arctic mountain, the sinew-tightening acid of hours under stress was taking its toll on me.

Finally, we could hold back for the rafts no longer. The kayak team made the decision to push on to where we felt we could usefully make that evening, and try to recce a decent camp. Within another two rapids, though, we hit an impenetrable wall. To the initiated it probably would not look as brutal as some of our other drops, but with Daz's expert eye, the drops and swirls came to life.

'It's death on a stick,' he said. 'Strainers, syphons and sumps everywhere. There is no line down through this, and if you get it wrong, everyone could die.'

James concurred. 'Yeah, you might bump through some of it and get lucky, but eventually you're going to get sucked in. Bad day. Let's call it.'

'But call what?' I asked. 'Where the hell are we going to go?' To either side of the river were steep cliffs and forest. You'd never carry a boat through it. And up above us looked like a rocky overhang, so we couldn't go that way. In fact, our only option was to bushwhack back upstream through the jungle.

As James and I hacked back riverside, I was suddenly stopped dead in my tracks.

'Jimbo!' I stammered. 'James, come and have a look at this.'

'What is it, Steve?' James responded, coming down to join me in the sand at the water's edge, 'Holy crap! Is that what I think it is?'

'Yes, it is,' I said, reverentially tracing the edges of the footprint with my fingertips.

'It's a tiger, right?'

'Not only is it a tiger,' I said, 'but it is a fully grown male tiger.' This I knew from my last Bhutan expedition. I'd spent over a month tracking tigers using their prints and poo, and had drawn on my hand the spans of tracks and what they meant.

'It's a male, and he was here since the last rains. When did it last rain?'

'Last night.'

'Yup, I thought so. He's been here today.' I then followed the utterly perfect deep track down into the water. 'And look, it's so fresh you can even see the print underwater in the sand. He went right down into the water.'

A little later, we were to find the fresh carcass of a Himalayan mountain goat the tiger had killed, no more than a stone's throw from where we were now. He had clearly come down to drink after eating his fill. But for now, our reverie was called away from the print by a shout in the distance.

Upstream, one of the rafts had again wrapped itself around a rock. This time, though, it was in trouble and contained not only some of our professional rafters, but also several of my crew, for whom this would be their first experience of being jammed inside a torrent with tonnes of water beating down on them. At the bank, the other rafters battled to throw ropes and drag gear to safety, trying to get our crew members out without harm.

Twenty minutes after the flooding, the raft was shaken free, but by then the darkness had descended. We could continue no further. With what energy we had left, everyone began to clear spaces on the riverside to put up their tents, 30 metres from the flies of the fresh tiger kill. We looked at the GPS. With the rafts, we had covered less than five hundred metres in the whole arduous day.

It was clear now to everyone that continuing with the rafts would be utter folly. It would take nearly a month at our current speed to continue the section to India with the rafts along. And continuing with just the kayak team, in isolation unmatched anywhere else on earth, with no chance of outside help? For another ten days or so? I knew I wasn't up to it. I felt like I'd suddenly been hit by a cricket bat. My energy had vaporized, and all I wanted to do was lie down in the bushes and sleep. Plus, it was clear from the state of my bowels that the amount of water I'd swallowed while being washing machined in the rapids had not been good for me.

Our little family gathered shivering around the campfire for a serious chat.

'Sorry to tell you, guys,' I said, 'but I'm squirting out my backside like a fire hydrant. I'm knackered, morale smashed. That's me done, I'm out.'

'Don't be sorry, Steve,' said Sal. 'We've all been there.'

I prepared myself to be told I'd let them all down, that it was going to be all my fault. After everything I've pushed myself through, this would be my failure to bear for ever.

'Well, I've had enough too,' said James. 'This river is just way too committing. I've never been on a river so relentless.'

'Yeah, this is going to go down as one of the most full-on rivers in the Himalayas,' added Daz. 'There's no shame in us saying right now it's beyond us for now.'

Chencho, who had been furiously chewing betel nut as if his life depended on it, now broke into a huge grin of relief, his red-black teeth exposed.

'Yeah,' he agreed, both thumbs up. 'Good!'

'I'm pretty tired,' said Sal. 'It takes it out of you, this kind of paddling. I'd definitely need a good rest. But if there was another 80 kilometres of this, it'd be brutal. We have no idea what we're getting ourselves into. Possibly unrunnable.'

'Maybe we can come back another time and finish it off,' said James.

Even in my pathetic state, after everything I'd been through, the idea of coming back here again with these amazing individuals was surprisingly exciting. After all, a huge part of this whole project was keeping the dream alive – we didn't want to be the team that finished all the new firsts, we wanted to be the team that proved there are still great firsts out there to accomplish! It wasn't about us leaving no stone unturned. And next time I could train really hard, get great at paddling, prepare properly, bring Helen and Logan to see this amazing country . . . but these were all dreams for another time.

Getting out from camp the following day was not easy, but was made considerably easier by Pema bringing his entire village down to help us. With my body having essentially given up, it was all I could do to hike out under my own steam, allowing a beaming 70-year-old woman to carry out my kayak strapped to her head.

We'd been out of any contact with the outside world for a few days, but the second we reached the high ridgeline above the rivers everyone's phones started buzzing with messages, and then there were grim faces all round. Tasmanian paddling legend Adrian 'Adey' Kiernan, who had led the rafting team on our attempt to make the first descent of the Baliem, had been lost in Nepal. He'd been attempting to run the legendary Humla, a gorge river very like our own, just a few hundred miles away. He was a staggeringly competent kayaker, so something must have gone very wrong indeed.

As this trip had proved, friendships are forged hard and fast on expeditions. We'd been five weeks on the Baliem together, sharing stories, drinking warm beer and rum, getting frightened and excited together. And just as Sal and James now felt closer to me than cousins, Adey had been a brother to me in Papua. I admired him immensely, and had really wanted to go out to Tasmania to smash some big trips with him. And I wasn't alone. The expedition paddling community is small and close. Everyone on the team knew Adey, liked him immensely and would miss him. And his passing made my brush with death all the more real. If someone with his talent and experience could be taken by the river, what right did someone like me have to survive?

It was a subdued party that hiked out the next day to the furthest reaches of the gorge, in order to find an appropriate ending to the expedition. We'd heard of a remote Buddhist chorten, a watchtower in feudal times, but now a rarely visited sacred site. The trail took us through an enchanted pine forest, the pathway a carpet of ankle-deep pine needles, spongey moss and old yak droppings. Every tree branch was hung with epiphytic Spanish moss, tendrils backlit gold with the low afternoon sun. The woodlands had a deadened silence about them, like a sound booth hung with heavy black curtains that suck up all sounds into their folds. An occasional thrush would burst into melancholy song, the notes all the more pure and powerful for the fact that the bird sang alone. Once in a while we'd chance upon standing stones, clearly erected at the trailside for auspicious return. There were stupas too, adorned with ancient faded, jaded prayer flags, their plaster cracked from generations of wind, snow and sun.

The watchtower itself stood on an open hillside, narrow and proud. The original roof had been replaced by tin, but in every other way it had not changed in centuries. A huge death crack ran up its brick sides – a memory of the earthquake that rent Nepal in 2015. Choughs and ravens swooped around its walls.

The attendant came out to greet us. He'd been there for

75 years, had lost his two children and wife in a fire there, and now chose to live alone, far from people. Despite this isolation, he greeted us warmly, smiling to reveal just one remaining tooth. I gave him the first thing that came to hand: a chewy sweet.

We asked his permission to put up some prayer flags, and he said yes. It was the perfect location. From our vantage point we could look down into our gorge far below. When the wind blew the right way, you even fancied the rumble of the rapids was audible. To the north, we could look all the way up the gorge towards Bumthang. Our height afforded us great vantage and a pleasing sense of quite how much we had accomplished. We genuinely had come a long, long way.

But then to the south, the mountains disappeared off into the distance. The closest ridgelines were dark green, then lighter green, dust grey, and finally just a blur. From where we were standing, it felt like the end of the earth; a frontier of such scale and scope that it is surely beyond my limited skills. We chatted idly amongst ourselves about whether the river would ever get completed, about what horrors, heartaches and wonders lay in its path. Perhaps there was value in leaving something undone, a piece of the puzzle unsolved, a last treasure for other adventurers to covet. We talked about coming back one day, and how we'd want to return together, our little family reunited in this unique and special place.

Then we strung out our prayer flags. Their constant fluttering in the mountain breeze would carry our hopes and wishes far and wide, round about the pendant world, blessing everyone and everything they touched. And without saying anything, every one of us wandered off to be alone with our thoughts for a little while. There was so much to be thankful for. The expedition had been an emotional one for everybody, but we all felt a little more alive, and a little luckier with that blessing. We said our secret goodbyes, a thank you to the sky, to the mountains and to our Dragon River.

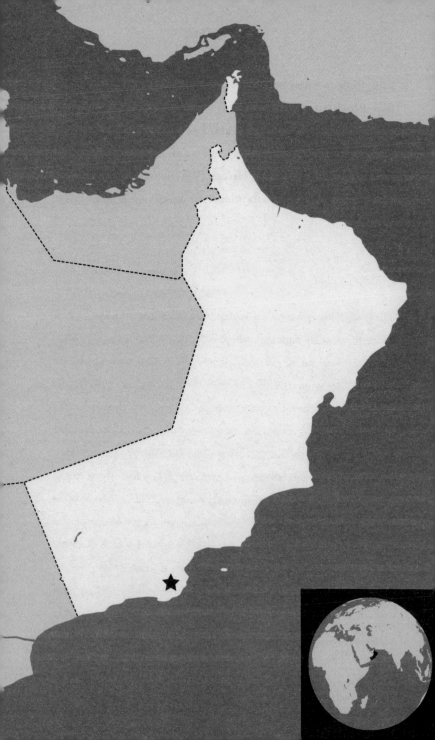

9. DESERT FORTRESS

OMAN

The grimy chimney has me hemmed in, dusty crumbling walls beating back the attentions of my fingertips and tippy toes. Far below, through billowing mist, I can make out the distant silhouette of Aldo, aiming his camera up at me. 'You're doing great, mate, stick with it!'

I teeter on my toes, legs spread wide across the gap, bridging the divide. Then with the palms of my hands I push outwards, edging upwards, trying to lean back against the wall to give myself purchase. My fingernails scrabble in the grit for something, anything, to hang onto. It feels like they are raking down a blackboard, which would usually make me shudder. Now it is the least of my worries. I breathe out, closing my eyes for a second, trying to steady myself. The wall in front of me is covered in enticing holds, but none of them is secure. If I take hold they'll splinter like cracker bread, sending me tumbling into space. Swivelling on one foot, I reach out to the other wall with one speculative toe, aiming to find my way out from my prison. My foot sticks, and I ease my weight on to it. Without warning, the whole block comes away, and a chunk the size of three housebricks falls, dropping straight to the ledge below, where Hazel is waiting. I don't even have the time to yell 'ROCK!'

The southeast of Oman is markedly different to Jebel Shams and the other mountains of the northwest. The region around the city of Salalah is blessed with an annual monsoon, bringing torrential rains that transform the desert. Though we were here in the supposedly dry winter months, there is markedly more vegetation: flowering pink desert rose blooming from bulbous roots, succulents, cactus and scrubby thorns, plus the weird dragon's blood bush, which weeps crimson blood sap when it's cut. Most definitive, though, are the flat-topped acacia trees, which make the escarpments and plains look more like sub-Saharan Africa than Arabia.

Our expedition was based in the Dhofar Mountains, dominated by the giant imposing escarpment of Jebel Samhan. Within just a few miles of the sea, the land soars into towering cliff faces 1,000 metres high, composed of 30-million-year-old limestone. These cliffs are vertical or overhanging for much of their length, and tower 500 metres high, with bulges and scars in orange, ochre and fawn, with ledges running across the faces longitudinally. They were known to be the last sanctuary for twinkle-toed desert wildlife: ibex, caracal and the critically endangered Arabian leopard. All these animals have been squeezed up into the most inaccessible places by human activity – the camel herders that wander right up to the lower slopes, the settlers making their homes down closer to the coast, the hunters who ravage the ravines in search of trophies and meat.

This would be the exact opposite of our last Oman trip, where we started at the top of Arabia's highest peaks before dropping down into unknown canyons. Instead, we would be beginning at the seashore of the Arabian Gulf, making our way ever upwards, fingers tracing the crack lines in teetering limestone.

When scoping the Jebel Samhan range looking for a possible first ascent, unclimbed peaks we could summit, new routes we could attempt, we had little to go on. There were no local climbing associations and no logs in the Alpine Club's literature of anyone ever having pioneered climbing here in the region. Getting more and more intrigued, we cast our net back to the 1970s, when British Special Forces were active in the area. Surely they would have done training climbs on this most iconic and obvious of features? Still nothing. In fact, we were astounded to find that no part of the escarpment had *ever* been climbed. And it was 44 miles, over 70 kilometres, long. It was literally a virgin mountain range. We would not just be attempting the slightly esoteric climbing challenge of putting up a new line or new route on the rock, we would be the first people ever to climb there.

There are many potential reasons for this. Firstly, it's remote and hard to get to. The logistics of even getting to the face were substantial, and would take time and effort. Secondly, despite the desert here receiving more rain than the rest of the Arabian Peninsula, it is still a desert. The escarpment points roughly south, meaning we would have no escape from the 40-degree scorch during the daytime. The closeness of the vertical slopes to the sea also meant that winds and moisture coming in from the water would inevitably smash the escarpment hard. There was no way of knowing what we'd get when on the climb, but there was real potential for very changeable and possibly horrific conditions. And then, of course, the fact that the rock was unknown posed its own challenges. We could start climbing at the base of the cliffs on solid, perfect rock, but who knew what we might find above us? It could be a lethal crumbly mess. I had already had more than my fair share of loose rock hell in this hard year. It was unlikely that my nerves could take any more.

The original plan had us coming here to climb directly after the Oman canyon trip, but we ran out of time. We readjusted to come here straight after the expedition in Bhutan, but it soon became clear that this was not going to work either. After what I'd been through on the Chamkhar Chu River, I needed to get home and have a long, extended cuddle with everyone in my family. Aldo and I were both battered, psychologically and physically, and were in no fit state to be attempting anything of this scale.

Aldo was nursing a forearm injury and hadn't been able to climb for months, I was still struggling to use my left shoulder and certainly had no time for an operation to set it right. I was overweight, out of condition and didn't recognize the haggard old face staring back at me from the mirror. It looked as if I hadn't slept in six months, which – funnily enough – I genuinely hadn't. It was clear that the whole team needed to take a break over Christmas. This stretched into two months, as I tour Australia every January (talking about animals and conservation).

Despite the fact that this in itself was going to be pretty full-on, I set myself the goal of getting in shape for Oman. Though two months felt like an awful lot of time in comparison to the build-up I'd had for all the other expeditions, it is actually a frighteningly small amount of time for what I needed to do: get myself in shape for a big wall climbing expedition that could be the hardest I'd ever done.

As I was going to be five weeks in Australia living out of a tour bus, I would have no time for any actual rock climbing, so I had to be creative about my preparation. Big wall climbing is all about finger and forearm strength, with stamina and strength-to-bodyweight ratio also being vital. So to start, I carried around one of those hand-strengthener clench-spring things executives in the 1980s used to use to make their handshakes stronger. Whatever I was doing, I'd make sure one hand would be pumping away, desperately trying to get Popeye forearms and shovel hands.

The only other thing I could do was focus on losing weight. If I could shift around eight kilos of fat, that would be equivalent to getting 10 or 15 per cent stronger. With the exception of a pig-out on Christmas Day, I basically lived on a diet of spinach smoothies and veggies, and trained at least twice a day, doing callisthenics and boxing for conditioning. I managed to get on actual rock a grand total of twice.

With just two days in between getting home from Australia and leaving for Oman, I packed my bags in a jetlagged daze, packing my Arctic sleeping bag for a trip to the desert, forgetting my toothbrush and any underwear, and only remembering that I would need climbing shoes and a harness when I was actually sitting in the taxi ready to head to the airport. The only slight positive was to look down at the bathroom scales and be able to see them, past what had two months previously been a burgeoning pot belly. I'd lost my eight kilos and could indeed crush walnuts between thumb and forefinger. Though that, sadly, would not bring me close to the league of my climbing companions.

Hazel Findlay had been my coach for a big wall climb in the Arctic in 2014. Hazel is probably chest height on me, with a calm reserve that hides phenomenal strength and skill. When she sheds her hoodie and starts taping up her fingers ready to climb, the muscles ripple across her back like cobras battling in a bag. Hazel was the first woman ever to climb the British grade E9, and is arguably the best female adventure climber our nation has ever produced.

John Arran, though, was to be our lead climber. John and I had done three expeditions together, and spent two months together in the field on big wall climbs. He is in his early-fifties but you would never know it; his energy and enthusiasm seem limitless. He's wiry, with mousy grey hair and atrocious dress sense, seeming only to wear the most garishly coloured gym kit at all times, no matter what the occasion. He looks a bit battered and weathered, but has arresting sky-blue eyes. Like Hazel, John also stands a head shorter than me, but I wouldn't take him on in an arm wrestle. He is the most formidably strong human I've ever met. His strength and talent are bolstered by a supernatural ability to climb on loose and lethal rock faces, where few mortals would ever dare to tread. John can scale a vertical cliff face with no visible holds as if he is walking up a ladder, and spreads his weight so efficiently that he can climb on unstable loose holds that any other person on earth would dislodge. And despite being borderline superhuman, he has a humility and self-deprecation that set him apart from anyone else in his field. Sharing so many exploratory climbs with him has been the biggest privilege of my life in expeditions.

Justin would also be back again after the joy of our last Oman expedition, and he would renew his bickering with Abdul Hamid – his local partner – on the mercifully short drives around the mountains. Aldo and Justin would have the monotonous job of hauling all of our hundreds of kilos of kit up the rock face, and keeping us stocked with water. They were joined by Miguel, an Australian wildlife cameraman and climber who was based in Oman and would be helping us with a variety of different jobs.

It felt like the *Avengers Assemble* of climbing superheroes. And then there was me, chuffed with myself for having used hand strengtheners for a month.

The drive-in to the base of the escarpment started down neat tarmacked roads, and involved dodging the hundreds of wandering camels in long shambling trains, who stopped in the middle of the road to chew the cud while eyeing us with disinterest. There are no actual wild camels in Arabia; every one is owned by someone. However, here in the Dhofar Mountains there is enough greenery for these animals to find natural forage in the same way as sheep may do in Snowdonia back home.

Our own train of white Hiluxes turned off the main road that runs parallel to the escarpment and made a beeline straight up towards the rock, bouncing up a boulder field that we were assured was actually an off-road trail. Alongside us ran a deep open wadi that anywhere else in Arabia would have been bone dry, but here ran with a trickle of water at its base. Around the flow was as green as an English country meadow. We jerked and jumped in our seats for the best part of an hour, rattling our retinas off and jangling the fillings out of our teeth. The escarpment never appeared to get any closer, which was a sure and uncomfortable sign that it was much, much bigger than it first appeared.

John had done a short recce to Jebel Samhan a couple of months previously, gazing up at the rock from a distance with binoculars. He had settled on a section of rock that he thought – albeit from a distance – would make the best possible climbing. It was as imposing as a giant's fortress in the clouds, with what appeared to be the gaping maw of a cave halfway up that we could sleep in. John had not, however, had the time to get close to this part of the rock face, or to recce a possible route in. Abdul Hamid had arranged with a group of local camel herders to bring steeds to help us transport all our gear to the rock face. When I mentioned my conservationist concerns about using animals to him, he laughed at me. 'That's not a problem here,' he told me. 'People here love their camels more than their children.'

This was undeniably true. When we reached the point where our 4x4s could progress no further, six camels stood chomping away, eyeing us with a total lack of interest. They didn't look massively happy about the prospect of carrying our stuff, but then camels rarely look happy about anything. The herders shared their disgust as we emptied our vehicles and unveiled all the ropes, cameras, tripods and water we needed. The head herder was Ali, an Omani of Bedouin descent with film star features, white beard and the suggestion of dark eyeliner around his black eyes. Ali was a showman, declaring his words with flourishing hand gestures, brandishing his sword about his head whenever the cameras swung his way and singing Arabic chants as he loaded his camels.

'Camels will carry 100 kilograms,' he pronounced, as if delivering a soliloquy.

'A hundred?' Aldo protested. 'You said 300!'

'Three hundred is on the flat,' said Ali. Then he pointed to the cliffs far above. 'We go up there.'

He had a point. All of us had planned for a three-hour hike on a well-worn trail. Instead, we were looking up at a near-invisible path snaking through rugged rocky foothills. The part of the escarpment John had suggested we climb was presently invisible, but it was pretty clear that it wasn't three hours' walk away.

'Well, what are we supposed to do with the rest of the gear?' Aldo asked.

'You carry or you leave,' Ali said. 'You look strong, I think you carry, yes?'

'It's half a tonne of kit, dude,' Aldo replied. 'I'm not that strong.'

'Don't they have you carrying big loads all day in the army?' Hazel asked. I am going to like her.

Yet another hour was spent streamlining our kit, getting rid of anything that might end up being extraneous. And then another hour packing and repacking our bags so that our human team all had as much as we could physically carry on our backs. I'd already packed my maximum load for a three-hour walk and was now *adding* a shedload of stuff for a death march that could take days. This was not going to be fun.

We'd all thought Ali was playing the game, and slightly conning us, but once we set off it was clear he'd been on the money. Camels are the ultimate desert ferry, but they're not designed for steep, rough terrain. They were deathly slow, and every ten minutes would want to stop to burp up some cud and have a good old chew. Terminally flatulent, they would additionally inflate their soft palate, drooping it like a huge blown-up dribbly pink balloon out the side of their mouth, while making a noise that sounded like a revving Harley-Davidson submerged in blancmange. They spit, bite and projectile poo at you, and are without question some of the most ill-tempered beasts to share a nice walk with. And this was not a nice walk. Doubtless it was beautiful with the ever-present majesty of the escarpment glowering over us, but the thorny acacia bushes and dusty heat made for a long slow slog. We crested every single ridgeline hoping to get the first glimpse of our climb, only to realize we weren't even close.

While my comrades spent the hike gazing up at the obviously ever-commanding presence of the rock above, and thinking up new ways of saying 'Oh my God, it looks scary,' I was finding the tiny things the most entrancing.

For me, probably the best thing about the walk was the grasshoppers. I know this sounds bonkers, but their astounding diversity and quantity were easily the most impressive I've seen. There were at least 30 species, many of them as long as my palm's width. Some were disguised as pebbles, round, sandy and dusty. Others were masters at flash coloration. The insects were well camouflaged and held their ground until you stepped too close. Then they leapt into the air and took flight using their concealed underwings, which were painted lurid reds and yellows. The purpose of this behaviour is to startle a predator, so it misses its chance to snatch a cryptic meal. Then, as the bug is airborne, the predator forms a 'search image' on the brightly coloured wings, but when the insect lands it folds those distracting colours away and is again camouflaged. The predator wastes time searching for the bright object it saw in the sky, leaving the bug to make good its escape. Stunning as these grasshoppers can be, though, they are also one of the most destructive natural forces in the world. When conditions are ideal (drought followed by rains and rapid vegetation growth), a chemical cue will change these individuals into gregarious swarming locusts that can wreak havoc.

The birds, too, were dazzling. The most common was Tristram's grackle, a black starling with vibrant red wing patches, incessantly vocal – especially when disturbed and in flight. The Yemen serin, endemic to this part of the world, sounded in voice a little bit like a big bunch of keys being jangled together. And then Bonelli's eagles, wing tips spread like fingers as they circled and soared aloft, searching for thermals.

By late afternoon the wind had picked up, bearing the dust about in swirling dervishes that peppered the cheeks and crusted the eyes like a churlish sleep sandman. I crested yet another false summit to come face to face with two young Omani camel herders, who had come up into the lower slopes to search for their lost camels. They were elated to see my expensive binoculars round my neck, and used them to stare up at the slopes. Both were barefoot despite the wicked acacia thorns, yet had spangly new iPhones which they used to take selfies with me.

The top of the escarpment was home to the area's telecom mast, visible no more than ten miles away, so we had full phone service for the rest of the expedition. The herders looked me up on Instagram and friended me on Facebook before leading me to the nearby water spring. Far from being a hidden gem trickling from the rocks, it was a concrete pool sitting on a shoulder, with a commanding view of the rock face, now tantalizingly near (though we still had not laid eyes on the bit we wanted to climb). As the spring was situated by a decent area of flat land, I decided we'd best make camp for the evening. The camels wouldn't make it for at least another hour, and it was clear we wouldn't be getting to the rock today.

While I waited for everyone else to arrive, I turned my attention to the rock face that was looming above me. Right above the spring, the rock appeared to be carved into the prow of a monumental dreadnought, with ledges running across it cutting the face into equal thirds. I traced several obvious lines up the face, all looking pretty achievable. There was plenty of dark orange rock that looked well weathered and as if it might not be totally lethal. To me, any of these lines looked like a perfect objective.

By the time the others had arrived, I was getting very excited about the rock above us and the idea of ditching the original route and cutting to the chase to climb one of the lines I'd spotted. To my chagrin, though, no one else seemed interested. John, who would see the walls of St Peter's Basilica and start looking for ways to climb it, lay down by the spring, looking away from the rock, pulled his cap down over his eyes and promptly fell asleep. Everyone else was focused on getting to our originally proposed climbing spot; it was going to take at least another full day, and we would then be way behind schedule.

'Do you want to have a look, Hazel?' I prompted, offering her the binoculars.

'You can't see it from here,' she responded, slumping back on her pack.

'No, at this bit of rock above us!' I said.

'I'm all right, thanks,' she said, delving in her bag for a healthy nut bar.

'It's kinda cool,' I pleaded. 'There are lots of what look like lines we could climb . . . I mean, not that I'm the expert . . .'

'What's that mountain in America with all the presidents' faces on it?' asked Aldo.

'Mount Rushmore,' Parker, our American soundie, responded.

'Well, this is like that,' Aldo said, 'but with arses.'

That caught everyone's attention. They all looked up.

'See that middle section?' Aldo gestured. 'You've got Stevo's massive arse there, then John's spindly little arse in the middle, then Hazel's arse over to the right.'

'So which crack are we climbing?' I asked. The giggles woke John up. He'd not slept the previous night and was exhausted.

'What do you think, John?' I coaxed. 'We won't make it to our intended route for at least another day. If we changed to here, we'd start climbing tomorrow. I mean, it looks to me like there's a natural line up through that chimney to the ledge, and then beyond that, we'd pick a route through the giant buttocks!'

John reached out for the binoculars. It felt like that moment where the man from Del Monte gets handed an orange. 'There's an obvious line up through the chimney there,' he said. 'Just your kind of climbing, Steve!'

Now Hazel picked up on the enthusiasm. She was desperate to get onto the rock, and cutting to the chase was certainly appealing for her. 'We should get up there and have a look,' she said. 'If we went that way, we'd be climbing by tomorrow afternoon!'

With the shadows lengthening we dumped all our bags, and the climbers sprinted up towards the new wall, while Aldo and Justin yomped off over the path to try to get to our original cliff face.

By the time Hazel, John and I were within reach of the rock, our minds were made up. A flattish field the size of a footie pitch just half an hour from where we would start climbing, with a giant overhung boulder shaped like a breaking tsunami providing some shelter, would make a superb basecamp, and the camels would be able to reach it. However, from there, it was an awkward and potentially dangerous scramble to get to the actual rock face, which we'd have to do multiple times, carrying heavy loads. The closer we got, the more intimidating the face became. Even Hazel was looking up at it now with trepidation. Big, overhanging, it looked like trouble. It had the air of how a toddler with a crayon would draw a medieval fortress; as if there should be armoured men behind every slit in the stone, poised with drawn bows and arrows, ready to unleash.

Once we reached the cliffs, though, it looked incredible for climbing. The limestone was like grabbing a hold of a piece of pumice stone, incredibly grippy, super positive to the fingertips and likely to provide great traction. The downside was that any rock that didn't receive direct weathering flaked off to the touch in slivery sheets. It wouldn't have held the weight of a resting grasshopper, let alone the toes of a hefty climber weighing a tenth of a tonne (i.e. me).

As we stood contemplating the cliffs, trying to rediscover some excitement to power-hose away the dread, the sun began to set through the drapery of cloud hanging over the escarpment. The sheer rock cliffs and cloud strata sliced up the sunbeams into rays, which cut down through the valleys, gorges and gulches all the way to the shining gold shield of the sea. Red-winged grackles and Yemenese serin flitted in tiny dark outlines through the glimmering yellow light, little monochrome pixies in an oil landscape that even Gauguin would have torn up as too garish and gaudy.

Nothing beats the desert at dusk. The scorching heat of the afternoon is still fresh enough in the mind that this pleasant warmth now feels like ice on a burn. It was one of those many moments we'd had in this year when nature suddenly jerks you back into real time, reminding us all of the pure privilege of being here. For days now, we'd all been so bogged down with logistics, filming and fear that we'd not taken even a second to just be – to stare out at the extraordinary vista and revel in its majesty, to enjoy the honour of what we were here to do.

The escarpment stretched off far beyond what the eye could see to east and west. It is as commanding a geological feature as you will see anywhere in the world, and our team were about to become the first human beings ever to climb it. Much as in Bhutan with our Dragon River, no one else would ever have that privilege again. You can only be the first once.

Tempting as it was to just stand and bask in the glory of the sunset, there would not be much more light and none of us had brought torches, so we scuttled off through the thorns and rolling rocks to make it back to the spring before dark. I jogged back with my mind racing, thinking of how we'd get half a tonne of kit by manpower up to the rock, working out how much water we'd need and planning how I'd describe it all to Helen.

I trotted in to spring camp just as Justin and Aldo returned. Their recce heading round to our original route had been as calamitous as ours had been triumphant. Any paths heading in the direction of our objective had been washed away by flooding, and it would probably take us at least two more days to get to our intended destination. The camels couldn't make it, so we'd need to do that trip there and back perhaps four times. By which time, Hazel would have lost patience and solo climbed the whole thing on her own. That decided it – the whole team would now turn their attention to our new objective.

That night we sat round a fire with our camel herders, drinking tea so sweet it sizzled the enamel off your teeth. I was overjoyed to see they'd boiled up some real rice, with plenty enough for me to eat. No dehy food for me tonight! That joy was short-lived when I found lumps of grey meat hiding in the grains: it was camel. As soon as we arrived, one herder (dressed for the catwalk in white blouse and black flowing pants and headscarf) had busied himself with draping pieces of fresh camel meat all over a leafless tree by their campfire, turning it into a biltong bush. That meat had then been added to the pot. Sighing, I returned to the kit bag and added water to a boil-in-the-bag nettle curry. It tasted like nettles, and for the next two days I was farting like a bison.

There was a lot of work to be done before we could even think about climbing. We needed to get all the climbing kit and ropes up to the base of the face, and ideally we wanted to be on the rock before it got too hot. This meant getting started while it was still dark. It was an early rise for all. We'd hoped to fill our water bottles and slog straight up for the cliffs, but woke to find all our camels troughing face-first down in the little spring, with camel spit forming an appetizing scum on the surface. Regrettably, we would have to filter all our water; particularly because many Arabian camels carry a disease called MERS (Middle Eastern Respiratory Syndrome), which is very similar to SARS. For the rest of the trip, if anyone so much as coughed, someone would mutter 'MERS' under their breath, and we'd all subtly shuffle away to drink our coffee alone.

My tummy was gurgling. And not just because the camels were putting me off my breakfast. Today would be the first day I actually had to climb. Having had a really good look at the route, it was clear it would be hard from the get-go, and would just get worse and worse as the expedition progressed.

The top section (hereafter known as the 'Headwall') would be sustained and probably overhanging throughout. This looked as if it was about 100 metres in height. Below that was a system of ledges, on which we could see vegetation and plenty of features. It seemed likely we would be able to walk about there.

Next down was another band of rock, perhaps 150 metres in height. This is where the 'buttocks' were located in our mental map of the route. At the base of the bums was a ledge running just below the butt clefts. Though this was not big, we all expected there to be enough space for us to sleep there. If we could progress through the first rock band today, then this would be our first ledge camp.

While I favoured following this ledge as far right as possible to find an easier line up, John and Hazel were being more ambitious.

The route Hazel was excited about from our first ledge camp would take us through a series of overhanging roofs that I could see from the ground was way, way out of my league.

Backpacks filled to bursting, we set off on the first of many scrambles we'd make over the next two days, picking our way past a vertical rock band and scrabbling over loose rocks to the bottom. We stashed everything beneath a small overhang, before beginning the pre-climb procedures we all have to get ourselves ready. The first pitch looked easy, but we prepared ourselves as if we were about to go toe-to-toe with Anthony Joshua, taping up fingers, adjusting harnesses, checking every cam and sling for easy movement and strength. The second we stepped off the ground we would be in a privileged world of heightened perception, a place where the hyperbole of the 'normal' world is replaced by the certainty of simple truths. When you climb as a pair, one leads and the other holds their rope. On an adventure climb like this, you literally hold your partner's life in your hands, and then they take your life in theirs. It's a covenant of trust with the most direct of consequences.

There was no certainty of what we would find ahead, but the giant boulders that littered the acacia slopes gave us fair indication. Much of the exposed rock that received direct weathering was as perfect as you could ever imagine. The holds were as carved, obvious and positive as you would find at an indoor climbing wall. Pieces of rock that had not been subject to erosion, however, were what we would describe as 'choss': crumbly, flaky and poorly connected to the main matrix of the rock. It would fall apart if you so much as breathed on it. We knew there would definitely be rock-fall, and that this could kill anyone waiting below. Even on sections that appeared easy we would need our attention to be laser-focused for every single second.

So these moments of preparation were critical. We'd be wearing gremlin cams on our shoulders,* which would unquestionably get in the way and make the already committing climb even more hazardous. Hazel answered questions tersely, resenting the delay and interference with her mental preparation. And with reason – the first lead ever on this rock was to be hers.

On a climb like this, the majority of the risk and the challenge lies with the leader. They need to do the route-finding, to read the rock as if deciphering a mystery. And due to the nature of protection, a fall for the leader can be serious, even lethal, whereas for the second, it's usually just a nervous but harmless rope swing.

Adventure climbing is broken down into pitches, a measure in climbing that breaks a rock face up into segments. These can be anything up to the full length of the ropes that you're carrying, which in our case was 60 metres. Carrying a 'rack' of 'gear' that would be slid, crammed or cammed into holes and crevices in the rock, Hazel would follow natural crack lines, placing this gear into the holes.

Her route followed a neat corner where two walls met, and led up into an alcove above. She cruised up the rock, waltzing up easy

---

* A new development for these expedition films – a camera system on a bendable arm that watches your every move, sitting on your shoulder like a gremlin!

terrain to find a spot to make a 'belay' – a safe stance where you can fix the ropes and bring up the following climbers. A good belay spot has ground where the climbers can stand and move around a little bit, and critically will have enough cracks in the rock that the leader can put in lots of gear, so they can make a system that spreads any weight on the rope, to provide support for the person climbing up behind them. A bad belay has three people together hanging in space, relying on a couple of pieces of gear held in on dust and prayers. For a heavy climber like me, a big fall would put a tonne of weight and strain through every piece of gear and equipment. There was no margin for error.

Hazel had disappeared out of sight into the alcove above, but we could still hear her clearly. She was spitting worse than the camels, and cursing fit to curdle milk.

'ROCK!' The call echoed round the walls. We leapt out the way and ducked under overhangs as great dinner plates showered down on us, smashing through the acacias and exploding on the ground.

'There's no gear!' Hazel shouted down (she meant no good holes and cracks to put gear into), 'and everything here is loose!'

'ROCK!' We all ducked back in as the hail descended, every piece substantial enough to shatter a helmet and the head inside it. For the next 20 minutes, John and I crouched, chewing on trail mix and filling up our stomachs with water so we wouldn't have to climb carrying any. We'd both been well enough prepared to put our sunblock on as soon as we'd got out of our sleeping bags when it was still freezing cold – if you put it on just before you climb it's impossible to get the grease off your fingers, and inevitably the sweat makes it run into your eyes.

It was mid-morning by the time I started climbing, the sun searing and reflecting off the rock like one of those contoured mirrors posh people in Cannes use to intensify their face tans. There was, though, a thick haze starting to form in the air, which seemed to be gathering just around us at the coalface of the escarpment. By the time I'd reached Hazel's crummy cave, that haze had turned into a billowing fog, and far from needing sunblock, I was starting to wish I'd worn a cardie. It was immediately obvious why Hazel had been so stressed while trying to make the belay. The open cubbyhole had a sloping floor that was covered in loose rock, with pieces shaped like tiles, matchboxes and hardback novels. You could not place a single tootsie on the floor of the recess without dislodging something, which would slide down and pitch off into space. It was lethal.

As John climbed up, Hazel and I played a high-stakes game of human statues, knowing that even a twitch of a toe would send something sliding, and John would be right in the firing line. When he reached us, John took the extra gear from Hazel and headed for the back of the alcove, where the next pitch would begin. I'd been trying not to look at it, as it appeared at first sight to be really difficult. The ceiling of the cubby had only one break in it – a jutting-out chimney with hugely overhanging roofs to either side. It looked technical, and right at the limit of the grade that I can climb. But that was if the rock was in good shape. If the rock was bad, I was going to be in a world of hurt.

The start was not auspicious. John pinged off the first hand-hold, and the second. And then, in order to get himself up into the chimney, he had to contort himself into a bizarre yoga position, legs akimbo, before twisting 180 and taking up the same position facing outwards. Any small nubbins of rock that he tested out came away at the slightest touch. It looked desperate. Soon he was completely out of sight, immersed inside the chimney, but every new move sent a shower of pebbles and dust down and out the bottom. It was like an old Victorian fireplace being cleaned, with soot shooting down from the flue above.

'I don't think you should climb this,' Hazel said. I had to admit, the same thing had very much occurred to me.

'I mean, look at the tag line,' she added, referring to the thin rope that the lead climber trails, so they can haul up anything extra they might need. 'This is really overhanging. If you fall off, you'll be dangling in space. And if you pull something off, it'll come down on me. Seriously, Steve, it's really dangerous.'

My mind swirled. I'd been pretty sure that I wouldn't be leading much on this expedition, and that it was possible I might have to skip some pitches altogether. But not on the first day! We'd only just got started! Yet if it was going to be dangerous to my fellow climbers, then I could not in good conscience try it. Hazel is a professional climber, she only climbs with the very best. It would be horribly unfair for me to risk her safety just so I could play my part in the first ascent.

I was freezing cold, my teeth were chattering and I'd wrapped myself up in the foetal position, trying to do everything to keep warm. My legs were also crossed as I was nursing all that excess fluid I'd downed to stave off desert dehydration. I didn't think Hazel and I knew each other well enough for me to unleash a couple of pints while harnessed to her on the little belay ledge.

Just when I made my decision to yell up to John that I was going to bail, he shouted first: 'You're going to love this, Steve, it's right up your street!'

Hazel lowered her eyes with a slight shake of the head. 'But what if he comes off, John? He'll be hanging in space!'

'Well, he better not come off then!' John shouted down. At times like this, I both loved and hated him in equal measure.

'Is it in my grade, John?' I shouted up.

'Well, it's spicy,' he called back down, 'maybe E3. But I've seen you climb harder. I think you should give it a go.'

'He says everything's E3,' Hazel muttered. I was torn now. I didn't want to be reckless and endanger Hazel, but on the other hand, I'd come here to climb. I'd regret it if I backed out now.

Gingerly I stood up. 'Take in, John, I'm climbing.'

Every footstep sent little skittles of rock skittering down the slope and off into the abyss. By now we were totally sealed in by a barricade of CLAG (Cloud Low Aircraft Grounded) and couldn't see more than a few metres. The fog deadened all sound and had turned our cubbyhole into a strange ethereal prison. I got to the corner, prised myself up off the ground and started to climb.

It turns out not only does John have supernatural ability with his own climbing, but he even knows my climbing better than me. While my bodyweight plays against me on fiddly climbing with little holds, my flexibility from a lifetime of martial arts comes into its own in a chimney. Bridging moves where you stretch each foot across a chasm allow you to take the weight off your suffering fingers.

'Nice move, Steve!' Hazel said, surprised.

I twisted to face out just as John had done, trying to spread my weight between feet and fingers. My left foot sought out a slightly proud nubbin on the opposite wall.

'Are you sure about that one?' Hazel was tentative. Pop! The rock disconnected from the wall, sending a lump about the size of three housebricks tumbling, to explode on the cave floor just metres away from her.

'Are you OK?' I panicked.

'I'm fine, you're doing great.'

I fought to get my breathing back under control. And then realized I hadn't fallen off! My main foothold had pinged, and I hadn't even loaded the rope. This was good. Maybe I did have this under control after all.

Squeezing myself back into the corner, I eased my way upwards, moving slowly, checking every hold before committing to it. It was nervy, but do-able. The only thing was I knew at some stage I would have to step out of the chimney, and I hated that. Anything in climbing that gives you a feeling of security is a serious psychological threat. Sooner or later you'll have to give that security up, and when you do, it's worse than starting again from scratch.

Halfway up, and that move had to be made. I turned sideways and onto the main wall. Just above me it bulged out, creating a short overhang. And I was instantly stuck. I scanned around on the blank wall in front of me . . . nothing. I raked my fingernails down the rock, chalk dust scraping deep beneath my claws.

'Oh crap!' I pronounced. 'I seem to be stuck.'

'What's the matter, big Steve?' John called down.

'I can't see what to do next, John.'

'Just make your way up a bit,' he said without even a hint of irony.

'Thanks, John. You should teach climbing for a living.'

I fussed a bit more, stroking the cool limestone as if it might come alive and flex into some nice clean holds for me.

'I'm really sorry, John,' I called up, 'I think this might be beyond me . . . Oooh, no . . . hang on.'

At that moment, I'd turned round and looked over my shoulder. I've been climbing for 30 years, but it still never fails to astound me how the most basic tenets are true. *Just work your feet up a bit and everything will change*, I was always taught. And it had. In moving out to the blank wall, I'd ascended no more than half a human's height, but in doing so, I'd completely changed the aspect of the wall behind me. Stepping back across to bridge again, the impossible suddenly became simple.

I progressed like this all the way up the pitch, stepping backwards and forwards across the void, inching my way up steadily. When it comes down to it, the seemingly superhuman attributes of someone like John can often be explained by their ability to just do the simple things perfectly. Again and again, in every situation, no matter how nerve-wracking. Before long, John's piercing blue eyes were gazing down at me through a slice in the rock. He was swirled in dense fog and wrapped up in a skin-tight hoodie in a shade of lime green no other human could pull off.

Seconds later, I was standing on the ledge alongside John and surveying our new home. It was simply incredible. The shelf itself was five metres across, with a roof that started at the back at knee height, then opened out towards the drop off. The floor was relatively flat red sand strewn with rocks and absolutely laced with footprints and droppings. On closer inspection, the cloven hoofprints belonged to ibex and rock hyrax, the former a startling mountain goat/antelope, the males of which have colossal curved horns, and the latter a little hamstery round bundle of brown fur, which thrives on vertical cliff faces. (The most popular fact known about rock hyraxes is that their closest living relatives are the manatee and the elephant!) There were also, though, some mystery mammal prints – potentially mongoose, though I couldn't be sure.

The ledge stretched off to either side a really good distance, offering not only plenty of places for the whole team to sleep, but also lots of opportunity for a good explore. There was no time for that now, however. We had little light left in the day, certainly not enough to spend time hoisting up our sleeping stuff. The thick fog had soaked all of us to the skin, and John and I shivered together with our T-shirts stuck to goosepimply flesh, looking like a pair of vacuum-packed plucked turkeys. Hazel was soon with us, having found the epic chimney of doom disappointingly easy.

'What did you think of the grade on that, Hazel?' I asked, gently prodding for a response about how tough it was.

'No idea,' she said. 'I never normally climb anything in grades that easy.' I decided I wouldn't ask again.

The number one concern now was to get a fixed line in place that we would all be able to climb the following morning. This would enable us to circumvent the pitches we'd done today, climb straight up the vertical rope and begin the following day from this highpoint. That done, we abseiled off into the fog and bombed down to basecamp in search of anything to wrap round ourselves to get warm again. My lousy packing meant I didn't have a jacket (seriously!), but my Arctic sleeping bag would come in useful. That night, everyone else would shiver and shake in the soaking wind, while I curled up, snug as a bug.

Before we could turn in, though, there was one more job to do. Miguel and our national parks guide Khalid had put camera traps out on the area around the cliff ledges and had them in place for several weeks before we arrived. While we had been climbing our first few pitches, they had slogged off around basecamp and picked up the camera traps. They'd been waiting on tenterhooks for me to arrive back, so they could finally watch them and see what animals were wandering around the fortress we were now calling home.

The first few films were innocuous enough. There were dozens of shots of Arabian partridge, clucking about in comedic fashion in front of the camera. A train of three honey badgers nose-to-tail beetled into the frame, noses down. (Honey badger don't care.) A rare Blanford's fox snuffled all over the lens and sprayed it down with wee to scent mark it. But this was nothing. The next image had us all cheering as if our team had just scored the winning goal at Wembley.

The camera was triggered by movement just out of shot. Then the scenery came alive, dark shadows and light patches transforming into spots, feline shoulders and arched spine. It was a female leopard, with golden eyes, and in stunning condition. The subspecies of leopard found in this region is the Arabian leopard, and is critically endangered. There may be as few as 30 of them left in these mountains. And we had two: a male and a female. On our camera traps we watched transfixed as the male followed the female up and down the ledge paths, sniffing where she'd been, making scent marks and scrapes as a signal to her and a warning to other potential suitors to back off and keep their distance. It was simply stunning. The most recent images were that very morning. The leopards had both walked the same trails just hours before us.

The next morning, we were up with the dawn. The night had been freezing cold, and there were a lot of drawn, haggard faces as we brewed up our porridge; it seemed I was the only person who'd actually got any sleep at all. Worse for the folk heading up the wall, the thick dense fog still hung over us, as if we were inside a storm cloud. Those whose sleeping bags were soaking wet would have no way to dry them and would be starting ledge life knowing they'd always be soggy, cold and sleepless.

We loaded up with as much weight as we could carry, and set off up the awkward ground. John had already climbed up the ropes by the time I reached the rock, and Hazel was next to climb. I've never seen anyone go up a rope so fast; with the fog cutting our sightlines down to as far as you could throw a stone, she was out of vision in less than a minute. I followed soon after, and by noon the three of us plus Keith the cameraman were up on our ledge below the buttocks, and ready to start the first pitch of the day.

On closer inspection, Hazel's chosen line out from the ledge looked near enough impossible, and instead she and John had decided to scout around to see if they could find alternatives. As we scrabbled round to the right, it started to rain. Not the damp, drizzling pea soup we'd had since being on the face, but actual rain. We were wet through when we reached the obvious crack where we hoped to begin. Although today we were a bit better prepared, there was no getting away from the fact that the day up there on the rock was going to be plain miserable.

Hazel started up first, and cruised an easy slab, before popping over the top and disappearing. She dragged out the entire rope in her search, 'questing' forwards to try to find an easy break in the rock face ahead. Eventually she called back on the radio: 'I've got as far as I can for now and made a belay,' she said, 'but I don't think that Steve ought to try and climb it. It looks like it could get spicy above here.' I decided to give it a go anyway and follow up the easy rock. But as soon as I got onto it, I faltered. The limestone was slick with rain, and my feet and fingers couldn't find any traction. With my nerves jangling, I stepped right from the blank rock, and into the big obvious crack feature. The air was heavy with the distinctive smell of bat droppings, and the surface felt as if it had been squirted with Fairy Liquid.

'That's an interesting move,' John said, with a heavy air of questioning. And then, just in case I hadn't got the message: 'There's not many people would think that was a better way to go.'

This was the easiest piece of rock I'd stepped on in years, but I felt insecure, like I might slip off at any second. And then my mind started racing. It was already midday. We needed to reach the slopes up above us today, or we'd be in trouble and unlikely to make our ascent on schedule. Surely the sensible thing to do would be to let them dash on, get the ground covered, and give them a chance to get us back on track.

'Tell her I'm coming down, John,' I said.

John was surprised. 'Really?'

'Yeah, don't you think it would be better if the two of you cracked on together?'

'No', he mused, 'but then I wouldn't have expected you to struggle on this section either . . .'

'It's going to be cold and wet and potentially dangerous. I'll just slow you guys down.'

'Doesn't seem to be much point without you along,' he said.

'This is your expedition after all. And this could be the easy day – I thought you'd be leading today.'

I waited for the customary surge of indignation and pride, for the hot flush of fury to burn, forcing me to reconsider, as it had done on every other expedition I've done in the last quarter of a century. But it didn't. I never felt even a buzz that I was making the wrong decision. As John wiped the grime off his climbing shoes and set off into the clag, Keith and I wandered back round the ledge to where Justin and Aldo were hauling up the bags.

'That's the right call, Stevie,' Keith said. There was a lot behind those simple words.

The boys were hauling bags up into an open crack cave, the roof no more than a metre or so above the rough sand floor, but opening up near the lip where the rock dropped away beneath and soared into vast overhangs above. The fog was still so thick that you couldn't see any of that, though, so I spent a few hours scrabbling around through the caves to see what I could find. There were loads of animal footprints and dung throughout, showing that somehow these nimble creatures managed to find their way up here – potentially to stay safe from leopard.

But the most exciting thing was the fossils. There were perfect conch shells that looked as if they'd been taken off a beachside tourist stall. A clam looked like a perfect makeup compact, plus there were dozens of marine snails and crinoids or feather stars . . . and every one of them made of stone. Our cave was clearly a bedding plane – a place where layers of marine limestone of different ages met. The various sea beasties had been covered on the sea bed and over millennia had turned to stone. Then sandy sediment had covered them, then more limestone. Now the sand was weathering out, leaving an abundance of fossils that made it feel like high-altitude beachcombing.

Ever the team player, I decided to pinch the choicest sleeping spot for myself. I started scraping the sand away to make it into a nice flat bedding spot. Keith was filming the guys hauling up bags, when I shouted out to him: 'Keith, you've got to come and have a look at this.'

Everyone looked over to see me sitting transfixed in amongst a pile of black rocks that had been uncovered as the sands eroded away. I took one gently in my hands and held it up to the camera lens.

'What's that?' said Keith.

'It looks like a vertebra,' I replied. 'From a whale.' At that moment, Aldo came on the radio, guiding Justin on how to get the next bag up.

'What's going on up there?' he asked.

'Not much,' Justin said. 'The bag's a bit stuck, the weather's still totally clagged in, and Steve found a whale.' Everyone laughed.

But then I found another fossil that looked like a vertebra. And this one looked as if it had washed up on a beach that very morning. It was smaller, only about the size of a side plate, but it had all the definition to the lateral processes, the perfect hole in the centre where the spinal cord should go. A quick excavation in the sand within a few metres of it revealed 16 more stone shapes, up to the size of a dustbin lid. There was also a score of other weird fossils: ossified tricorn shapes, and a weird bone that looked like an African tribal mask that I couldn't place.

'But how do you know it's a whale?' Keith asked.

'Well, we're in 30-million-year-old marine limestone. It can only be a whale or proto whale,' I said.

'So did the sea level drop since then?' Aldo asked. That's what had happened with our caves in Mexico.

'No,' I said, looking at the altimeter on my watch. 'We're 1,400 metres above sea level here, the seas never rise or fall more than about 20 metres.'

'Plate movement then?' Aldo asked.

'Exactly.'

Over 30 million years ago, after the dinosaurs had been wiped out, but before even the most rudimentary of hominids appear in the fossil record, this little cave had been the sea bed. Countless sea beasts had breathed their last and sank to the bottom. Then, in the millions of years that followed, tectonic plate movement squeezed the sea bed and lifted all that rock up to where it is now: 1,400 metres above sea level, and part of a vertical cliff face towering over the desert.*

Just before last light, Hazel and John abseiled down through

---

* Upon returning from this expedition, we sent photos and videos to a number of noted academics to get their opinions. Several said they are more likely to be fossilised burrows from shrimp-like marine organisms. While I concur that the vertebra-like shapes could well be just that, the tricorn bone is still a mystery. Sadly it seems unlikely that we will be able to send anyone back to the cave to find the bones and clear things up once and for all!

thick fog, miraculously landing exactly where they'd started climbing that morning. They were soaked to the skin, wild-eyed, and Hazel was shaking with mild hypothermia. She described the abseil – ditching off into space with no idea of where she might end up – as being the scariest she'd ever done. While she crawled into her sleeping bag in all her clothes, John talked us through his day. They'd power-scrambled up sketchy rock, slithered around on greasy slabs and pushed the route on to the brink of the broad walkable ledge above. They'd succeeded in getting there, which had very much kept the expedition on track. But if I had come with them, there's no doubt it would have evolved into what is affectionately known as 'an epic'. Bottling out on the day's climbing was a very good call indeed.

For yet another night I was glad of my massive sleeping bag. Even though I'd made camp inside a womb-like cave, the air was so thick with moisture that everything was drenched. The bitter wind made it feel far colder than the five degrees it read on the thermometer, and I felt briefly sorry for the others, shivering away in their lightweight sleeping bags. Right up till the point I started snoring . . .

Around 2am I woke up to go for a wee, and stepped outside of my cave. The tendrils of mist were melting away. Stars twinkled above, and dim lights from the small settlements were visible leading out towards the coast. At sunrise, we woke to a glorious red burn, searing through what remained of the cloud.

As soon as I turned on my phone, a video came through from

Helen, of her doing a song and dance routine to Logan, and him chuckling with that baby giggle that has evolved solely for the purpose of breaking a distant daddy's heart. I recorded a little video to send back, showing the seemingly infinite wall of the escarpment, glowing peach and pear shades off into the distance, with the few fluffy clouds far below us and the castle of rock above. The melancholy took the shine off the perfection. I would have given anything to teleport them both here, to share this incredible high with them. They felt so very, very far away.

We had a little time that morning, so Aldo, Justin and I sat on the edge of the drop, sipping our coffee, tracing out the miniature marvels below. We could see our basecamp, where the film crew would surely be being roused by belching camels, fiddling with long lenses and drones, and by now working out how to get up to join us on the summit. Beyond them we pointed to the spring where we'd camped two nights before, and even further down to where we'd started our trek and our shiny white 4x4s, standing out like teeny Matchbox toys against the desert scrub.

We could also see to where our original climb would have been, and had to count our blessings we hadn't tried to get round there. The 'trail' headed for miles over treacherous steep terrain, which the camels could never have attempted. It would have taken us a week just to get there. And then we all looked up. Far above us were overhanging walls, then gargantuan black ceilings, glowering down on us with genuine menace. And hanging down, through the only possible climbable line, was one spindly-looking rope, and John jugging his way up it like a teeny spider on a thread of silk. It turned my tummy in knots just looking at it from below.

'So where did you sleep last night, Aldo?' I asked.

'Well, it was a new one for me,' he said. 'I slept inside a 30-million-year-old whale.'

He stared into his stodgy porridge. 'I'd cut off my hand for some fresh fruit.'

'Said no Scotsman ever,' I jibed.

'Weirdly,' said Justin, 'I just found these in one of the bags.' He pulled out a small sack with eight little oranges inside it.

'No. Fricking. Way,' said Aldo. 'I just carried those all the way up here.'

'Which I think means you own them, doesn't it?' Justin suggested.

'Standard hauling rules,' said Aldo.

Two minutes later, with orange nuggets dripping out of his beard, he declared it the finest orange he had ever eaten.

'What are we going to do with all this water?' Justin asked. We'd provisioned our water based on us climbing and hauling in burning sunshine and having to stave off dehydration. So far we'd been doing everything in freezing fog, and no one was drinking anything! We were hauling 120 litres of water up the face, and every litre is a kilo. It was by far our biggest weight issue.

'Well, we could tip it,' said Aldo. 'But then we'll get into trouble, get stuck on the face, the sun will come out and we'll all melt like Nutella.'

'And you can't tip away water in the desert,' I said. 'Surely that's like a crime or something?'

'Easy for you to say, Backshall, you're not hauling it up this chossheap,' said Aldo.

We decided we'd take it with us up to the headwall, before deciding what to do with it all.

The sun and clear skies didn't last for long, but enough time for everyone to dry out their sleeping bags and clothes, and feel like we'd hit 'system reset'. Today was to be a haul-athon. John and Hazel pushed on up to the headwall to try to scope out our route to the top; Justin, Aldo and Miguel got lines in place to haul all our gear up to the ledges, and I took the job of man-carrying all that gear up from the top of the haul lines to the base of the headwall. It was a half-hour steep uphill scramble, which was quite fun the first time. Much less so on the tenth trip, hand-carrying 30-kilo jerrycans of water. When John and Hazel started calling for specific loads on the radio, I ended up bobbing backwards and forwards like team gofer. It was a big, physical, heavy day for everyone, and by the time we'd got everything and everyone to the rock face, it was already dark.

That night, while I forced down my nettle curry, the silence was so powerful was almost ominous. The only sound that broke the roaring quiet was the eerie high-pitched warbling song of the Arabian scops owl, which echoed around the headwall like the song of a mourning ghost. The sky was a kind of clear you rarely see, every star and planet battling to be noticed beyond the smear of the Milky Way. We sat and pointed out Orion's belt, Venus and Mars. Some of them sounded made up. I'm pretty sure Aldo's 'The Giant Bell End' is not a real constellation.

The next morning, we woke to glorious sunshine. This was to be our final push to reach the top and we would be starting within metres of where I'd bedded down for the night. The route began with a little overhanging ceiling, with great big tasty-looking holds we could use to power up through it.

'Give me a spot, Steve,' John said, and jumped up onto the ceiling to give it a try. As soon as he was head height, he trusted his weight onto one nubbin of rock which practically exploded under his grip, and he ended up sprawled in my arms. This was not an auspicious start: John does not fall. However, not to be undone, we decided that this should indeed be our line. An hour later, we were slathered in sunblock, bellies slooshing with pre-hydration and fingers taped, ready to ascend. This time both Aldo and I 'spotted' John as he began up through the overhang, and yet again he popped off a big hold as he fought through the lip of the ceiling.

From here it seemed the climbing tended right into a crack that looked much easier and would hopefully take us way up above where we could see. However, once John had reached the crack, he fiddled around in it a bit before rejecting it as too loose, and headed back left onto what looked like much, much harder terrain.

Over the years climbing with him, I've started to decipher the hidden messages in his speech. If John says a piece of rock is 'fun', it actually means: 'this climbing is out of the realms of any normal mortal.' If he says a climb is 'interesting', it means: 'most professional climbers would run screaming and bury their head beneath their pillow.' And when John says something is 'exciting', it means: 'it was suicide even considering this, you are probably going to die.'

What I've learned to do instead is listen to the things John doesn't say directly. If he's prompting me to lead the climb, it's right at my limits but achievable. Telling me I 'should consider it' means I should not consider it. Then there's his breathing. Up above me right now, I could hear John puffing and blowing, clearly undergoing considerable effort and stress. And if he's working hard, then I'm going to be totally out of my league.

'This is really quite hard,' John puffed down to us, 'and there's lots of loose rock. It's really quite dangerous.'

Aldo and I exchanged glances. John might as well have asked: 'Have you all written out a will and got your funeral arrangements sorted?'

'What do you think, John?' I called up as he fixed the belay. 'Is it doable for me?'

He paused. 'It might be,' he said. 'Give it a go.'

My turn, and I pumped up through the ceiling, flexing and gurning, hoping I looked like Sly Stallone in *Cliffhanger*, but more likely Harold Lloyd hanging off the minute hand of a giant black and white clock. From there the going got easy for the next few metres, before bulging out into a dramatic 'tufa' feature, with grippy flowstone with excellent traction, but in this case lousy crumbly holds. Anyone who climbs will know that my role as the second was an order of magnitude less dangerous than that of the lead, but even so, there were some heart-in-mouth moments. The climbing was right at my limit; Aldo and Hazel were directly below me, and I really didn't want to kick anything big down on their heads. And then more than anything, I felt I had something to prove. I'd bottled out of one major section of climbing, and done nothing to earn the respect of my climbing companions. This was going to be my only opportunity to do that.

In physical climbing like this, your major enemy is 'pump'. The tiny muscles, ligaments and tendons of your fingers and bulk of your forearms can be taking your entire bodyweight in a way they never would in normal life. I was on that pitch for near half an hour, and my excessive bulk was held on my fingertips for the entire thing. Eventually, lactic acid builds up in the muscles; they become rock-hard and shivery and utterly useless. You couldn't hold on to an ice-cream cone, even if the flake was melting out. Two-thirds of the way up, and I was pumped.

'Come on, Steve,' Hazel shouted up. 'That's awesome!'

I forced myself to focus. It was actually quite awesome! Here I was, a vertical mile above the ocean, on a rock wall no one had ever climbed before. I was on a climbing team with two of the best in the world, and within sniffing distance of reaching the top and being a legit part of the first ascent team. I gritted my teeth, and stepped out over the crux and up towards John. I wasn't quitting today.

'You really can climb, can't you?!' Hazel shouted up, as I stepped up to the belay and gave John a massive hug. I might as well just have been knighted by the Queen!

It was fully five hours of climbing later that I clawed my way up the last few vertical metres to where John was sitting on the summit. It had been three pitches of varying difficulty, with the endless threat of rockfall. At one point I dislodged a boulder the size of a beach ball, which hit the ground a hundred metres out from the base of the headwall, showing just how overhanging the climb was. Another rock was much smaller, but impacted metres away from where Hazel sat below me, sheltering in a little cubbyhole. The rock smashed onto our ropes, and it was only later as we coiled them up that we realized it had cut through the outer sheath of the rope, exposing the white core.

As we stood together on the summit plateau, sipping sweet tea and gazing out to the setting sun (while Aldo and Justin sweated below, dodging falling boulders and hauling up water we would now certainly never drink), it seemed this had been a perfect expedition. It had something of everything: a genuine first of real significance, a flawless team, hard work, a bit of fear but not too much, glorious views and stunning wildlife. But more than anything, it had the unexpected. Not a single day had gone according to plan, and every second had been a leap into the unknown. That unpredictability is what makes these days of exploration so special. In the modern world, we are always seeking to control every aspect of our lives. We want to schedule and compartmentalize, to make rules and order in all things. A good expedition is the opposite of all that. Sure, you plan as much as you possibly can to begin with, but then you just let it happen. It is fractal chaos, glorious guesswork and a steaming dump all over the rulebook. It promises a mishmash of experiences, some miserable, some beyond brilliant, most totally unexpected. Because you never know what's coming next, you're forced to adapt and react, and your decisions on those conundrums play out in real time, and really matter. The outcome of your decisions can be the difference between success and failure, even in some cases the difference between life and death.

Perhaps what we are seeking most of all from adventure is self-determination, and control over our own lives. But for me, it's more about the ability to wake in the morning having no idea where you'll sleep that night, and that being OK. Crawling out of my little cave to see the first sunrise after the fog, I felt more like a Neanderthal than I ever have before, surveying my hunting grounds from a high vantage point – taping up my fingers to climb being the equivalent of sharpening a spear for the day ahead. Standing here on the summit plateau, watching the sky start to bleed crimson, the Egyptian vultures circling high above us and the mist pouring over the precipice like a waterfall, we felt like kings.

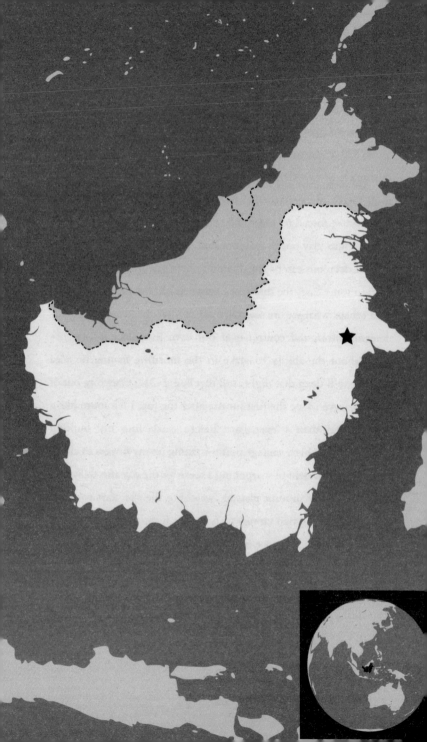

10. IT IS NEVER BROKEN

BORNEO

The vagrant's hand in front of me doesn't seem to belong to my body. Scratched, pickled and water-wrinkled like a corpse pulled from a sewer, knuckles swollen from bee stings, wrists bloodied with leech bites. Black fingernails pull on ragged limestone, but vines cling on, holding me back. Boots skitter in slime. Around my waist is a rope, but it's not attached to anything. If I fall, it will do precisely nothing to save me. I hook my elbow round the wrist-thick trunk of a scrawny sapling and take stock. Shadows are lengthening and black clouds are brewing. We are many hours from camp, and fewer hours from darkness. Half of the team are out of drinking water, and the humidity is brutal. The smart call is to turn back. But no more than metres above me, near the fang-like summit, is an unknown cave that could hold secret treasure. This is not the time for smart calls.

Sometimes in life things take on a pleasing symmetry. Like a good limerick ending with a neatly rhyming rude word. Or the way that *Expedition* drew to its conclusion. Our final destination was the jungles of Borneo. Exactly where I went on my very first solo travels before university in 1991. Here, I swung a machete and picked a leech off my privates for the very first time, back when I barely even needed to shave.

This final mission would take us to the southern portion of Borneo, which is part of Indonesia, where I had my first proper job.* All being well, we would finish this final voyage of exploration on my 46th birthday, and be home in the UK exactly a year after I had set off to explore the cenotes in Mexico. This Borneo trip would also feature underground exploration, with a high chance of finding ancient cave art, some of which could even be primitive hand prints, near identical to our Mexican 'small hands cave', despite being on the other side of the world.

Freaky, right? Too good to be true? Well, yes, I thought it kind of was. Actually, Aldo and I had been dreading this one. We both know Borneo well, and had sat glumly over more than one pint, talking about how awful the expedition was going to be. There were lots of reasons for our concerns. Borneo is the third-largest island in the world, and made up of three nations. To the north are the two Malaysian states of Sabah and Sarawak. In between them is the oil-rich sultanate of Brunei, which hit the international press

---

* This was writing a guidebook to Indonesia's wildest parts. In 1997, I also undertook my first proper expedition, which was to the Indonesian-occupied western half of New Guinea, then called Irian Jaya. This trip is recounted in my earlier book, *Looking for Adventure*.

the week we departed, as the Sultan had just declared homosexual relationships punishable by being stoned to death. The south of the island (making up about two-thirds) is the Indonesian state of Kalimantan, and for the most part is the Wild West, an unruly place of frontier-logging towns and transmigration, where urbanites from overcrowded Java have been shipped out to the jungle to find a new life.* We were bound for the east of Kalimantan, a region called Sangkulirang, which features arguably one of the most important limestone karst mountain ranges on earth. And is scheduled by the government to be smashed up to make concrete.

Sangkulirang is covered with impenetrable montane forest, as hollow as the Tinman's heart with caves. Bloody caves! It was coming towards the end of the rainy season. Even just to get to the heart of Sangkulirang to begin the trip we'd have to undertake huge heavy hikes, on arguably the worst terrain on earth,† in

---

* The highly controversial Indonesian and Dutch colonial policy of *transmigrasi* has taken mainly Javanese people and relocated them in the least populous parts of the archipelago, often with catastrophic consequences.

† For me, limestone karst, when covered with tropical rainforest, is the single most treacherous terrain. Aldo argues for mangroves because of the bugs and brackish water, but I reckon there's no contest. On the karst itself the vegetation is tangled and dense, with heaps of tearing spiky vines and splintering spines. The rock is simultaneously as slippery as if covered in oil, while also being so sharp and jagged that even a small slip can result in serious injury. As it is porous there is no surface water, so if it doesn't rain it's easy to run out of water (in the rainforest!).

driving rain, with jungle nasties eating us for breakfast, lunch, dinner and snacks in between. We were going to be wet, muddy and miserable. There was no guarantee of us finding anything spectacular. On previous expeditions here I've managed to progress just hundreds of metres *per day*. The forest is hard to escape, claustrophobic, and sometimes seems to be obliterating your will to live by insidiously growing mould between your toes and fungus on your soul.

Borneo is also the place on earth that makes me most glum when it comes to conservation. Flying in on my first ever visit in 1991, I can remember looking out of the plane's windows to the horizon in every direction and seeing nothing but jungle. In the verdant grottoes beneath I found frogs, geckos and snakes that can fly; monkeys with giant bosoms for noses, and tattooed locals with tales of cannibalism and headhunting. It was the place of my young explorer's dreams.

When I returned in 1997, it was during the year of the 'Great Burn', when so much of Borneo was on fire (mostly due to slash and burn farming and illegal logging) that a cloud of smoke hovered over the entirety of Southeast Asia. For 1.2 million square miles. On the streets of Singapore you could barely see across the street for months, daily visibility averaging 100m. An estimated 100,000 people lost their lives to respiratory problems in the aftermath.

Later, in 2005, I spent several months here for the BBC's *Expedition Borneo*. This time from the sky I saw below me out to the horizon . . . nothing but palm oil plantations. Uniform, undulating, geometric lines of oil palm trees. A pristine Bornean forest thrills to the sounds of singing gibbons, wailing cicadas, barking frogs. The plantations are silent.

Every few years I return, and my mood turns blacker and blacker. In 2016, Helen and I spent months fundraising with the World Land Trust to buy a corridor of precious forest alongside the Kinabatangan River in the north. On several fundraising drives we managed to raise nearly half a million pounds to buy forests that would then be protected for all time (I believe this to be one of the few truly effective means that we can use now to alleviate the assault on tropical forests). The jungle lots we purchased sit like tiny green squares amongst a vast blanket of yellow-brown deforested sludge. Year on year, more of the dwindling forest disappears, more of the coast is clogged with plastic rubbish, fewer elephants and orangutans cling to survival. Borneo's rhinos are functionally extinct, its coral reefs are dying, its skyscraper-sized trees ground to pulp.

I'm a positive person, who ardently believes that hope is essential, and that we should never give up on a place and its potential. I believe in empowering people to think that they can make a difference, to think that they can change the world. Borneo has challenged that view again and again. This has become the place that throws that philosophy back in my face.

Leaving Helen and Logan for this final trip was again brutal. He was now eight months old, and every day seemed to mark a new milestone. Just a week before I left, he crawled for the first time. Within 24 hours the entire house was on lockdown, with every power socket, glass vase, heavy object and sharp edge welded down with gaffer tape or stuffed into a locked cupboard. And then, with just days to go, he suddenly decided I was his favourite thing in the whole world. Every time he looked at me, he started to bounce up and down like Tigger, his eyes alight and a giant smile across his face. He started to be soothed to sleep by me at night, he gabbled 'Dadadada' non-stop. Helen and I had to be disciplined not to spend half our lives with our phones pointed at him, videoing every single precious second, not wanting to miss a single breath or heartbeat. And yet again I was off. Back to the jungle where I knew there would be no contact at all for at least three weeks.

I'd convinced myself it was going to be the worst trip ever. But as always with the expedition game, the only rule is that you never know what you're going to get.

My negativity had kept me from putting much effort into researching Borneo's wild places in recent years, believing them all to be living on borrowed time. However, in going back to studying satellite imagery of Borneo in depth, my eye was drawn to a dark green splodge across the centre of the island, spanning all of the three nations. A huge mountainous spine, with no roads, no villages, no evidence of logging. And when you look at it on a map of the whole region, you can see that this dark blot is the same shape and size as Java. An island that is home to 141 million people.

In 2007, the World Wide Fund for Nature brokered an agreement to protect the 'Heart of Borneo', in a proposed reserve of over 100,000 square miles. In the three years that followed the agreement, 123 new species were recorded there. Sangkulirang lies to the eastern tip of this remaining haven. Yes, it is surrounded by plantations and vulnerable to exploitation, but it still holds 3.7 million acres or 5,792 square miles of forested peaks. That's more than half the size of Belgium.*

As so much about this mission depends on anthropology, and the movement of peoples across Southeast Asia, it is essential to know something of the history of the region. Up until about 13,000 years ago, when our planet was still in the grip of the last ice age, Borneo was not an island. With so much water locked up in ice, the seas were roughly 120 metres lower than they are today, and Borneo formed the eastern tendril of the vast Eurasian continent. The land bridges enabled human beings to migrate out from their heartlands – originally in Eastern Africa, but then across the Middle East and Asia – to colonise every corner of our planet. *Homo sapiens* migrated to Asia between 80,000 to 60,000 years ago. Current thinking is

---

* Excuse this clumsy comparison – trying to put land areas into context is the bane of my life. Sangkulirang is actually the same size as Hawaii, but that comparison asks too many questions: just Hawaii, or all the islands? And is that with or without sea? And before you ask, the industry standard doesn't work in this case, as Sangkulirang is only 80 per cent of the size of Wales! For anyone interested, the entire Heart of Borneo park is the size of New Zealand.

that they reached places like Borneo, Papua and even Australia by around 40,000 years ago. The main evidence for this hypothesis is caves like the ones we were hoping to explore.

Throughout Indonesia, there are caves containing intriguingly similar cave paintings. The most common is the negative hand stencil, like the one we found in Mexico – though these are no more than 3,000 years old, a comparative blink in historical time. They were made by someone placing their hand on a wall, and blowing pigment over the top of it. (A positive print is where the hand is dipped in ink and pressed to the wall, though these are not found here.) Negative prints have been found in caves on every continent bar Antarctica. Indigenous peoples in Australia still use this as part of their own dazzling art today, while the Cueva de las Manos in Patagonia has more than 800 prints that date from 2,000 to 9,000 years ago. Negative prints dating from 30,000 years ago have been seen in European caves. However, those in eastern Kaliman-tan are older still. Much, much older, and they have led to a greater understanding of how human beings expanded across the globe from their origins in Africa.

Attempts to date the prints using carbon-14 dating drew a blank. Carbon-14 is only accurate to 5,730 years, and it was clear these paintings were far older. Also, the paints used were hematite – almost pure iron oxide containing no organic matter. Later methods involved dating the translucent curtain of flowstone that has formed over the top of the paintings, assessing the presence of uranium as it decays into thorium. The results were staggering. Some of the paintings in Sangkulirang were near 52,000 years old. This is 12,000 years before scientists previously believed human beings were there at all.

Cave art also tells a tale about the people that lived here. Creative processes like these are evidence of advanced intellectual processes, of ritual life, of awareness of the world and its representation in symbolic expression. In short, of advanced culture. There is an argument that etchings and paintings are more significant than stone tools or iron blades, as they are a sign that the people that lived or passed through here had culture, and time. They were not preoccupied every hour of every day with finding food and defending themselves, but had leisure to grind up paints, to experiment with hues and pigments, to make and value art.

Our secret weapon here was Pak Pindi Setiawan, the global expert on the rock art of Sangkulirang, and a legend in these parts. He is a professor at the university of Bandung in Java, but in the mid-1990s was brought here for the first time by a pioneering team of French cave explorers. They'd seen art in the mountain caves and needed some professional advice. What Pindi discovered has transformed the way we think about human expansion across the globe. Despite his great achievements, he is a terrifically humble man. Also, despite being a city academic, he is effortlessly at home in the forest, trundling through all the long death marches with ease, making camp whenever and wherever we find ourselves. We were reminded for the umpteenth time of how this year would have been impossible without experts like him.

Just as this region is critical in tracing the migration of the human species, it is also fascinating in biogeographical terms. In fact, biogeography, which studies species distribution across geographical areas, is itself a science that evolved from the work pioneering naturalist Alfred Russel Wallace did in this very region, which is often known as Wallacea in his honour. To the east of Borneo runs the legendary Wallace's line: a deep sea trench that was a channel even during the Last Glacial Maximum, when the most amount of global water was bound up in ice, and sea levels were 120m lower. To the west of it, the land was all connected to the Eurasian continent, to the east on the other side of the deep channel it was joined to Australasia.

Wallace made the observation that to the west you found tigers, leopards, rhino, elephants, but there was an abrupt, invisible boundary, beyond which they did not occur. To the east of his hypothesized line you found the kind of animals you'd expect to see in Australia: kangaroos, other marsupials and giant lizards. Islands like Java and Borneo could be hell and gone from the mainland, but still broadly shared the same wildlife. And yet, either side of the line, islands could be a metaphorical stone's throw apart (like Bali and Lombok) and yet their fauna was from different continents. This simple idea was key to unravelling the mystery of how tectonic plates are formed and move, and how our planet's oceans and landmasses have changed over time.

After landing in the large town of Samarinda, the drive north to Sangkulirang took two days. Two days of post-apocalyptic landscapes

I will never forget. As recently as the early 2000s, this entire distance would have been covered in primary forest. Now the paved highway carves through a wasteland. Open-cast coalmines are surrounded by miles and miles of orange mud. Burnt-out tree skeletons stand like tortured scarecrows amongst mud hills all the way to the horizon. It was actually a relief to see palm oil plantations; at least there was something growing, something green. The sides of the highway were littered with smouldering piles of plastic garbage, and rotting bundles of palm oil nuts.* It filled me with utter despair. So I'd close my eyes for an hour, allowing the jetlag to mess with me and the roll of the bus to take me away. When I opened them it was always to the same sight.

Finally, we came to a halt at a frontier *kampung* (village), with a single row of rickety wooden huts on each side. Those to the right bordered the road, and then dropped away down to the river, where our expedition would begin. The waterway was in spate, milk chocolate brown, washed with tree trunks and boughs. The *kampung* clearly existed only as a trading post for plundering the riches

---

* Eyes on the Forest and the WWF list Kalimantan as suffering the second most cataclysmic deforestation of any forest anywhere on earth. The worst is Sumatra (also in Indonesia). While our increasing use of palm oil is portrayed as the demon, market forces and political short-sightedness are the bigger issues. The uncomfortable truth is we could outlaw palm oil tomorrow, and deforestation would continue at the same rate, for crops like acacia, soy and sugar cane. This is why I'm so convinced that forest purchase and protection is the most immediate and pragmatic solution.

of the forest beyond. The house porches were hung with cages full of songbirds. A sad-looking baby monkey pulled at the metal chain round his neck. On one roof was a skull of a crocodile, on another the skull of a cat, probably a clouded leopard. Piles of freshly cut planks sat on even bigger piles of scarlet sawdust. Several of the giant trucks that were thundering past (laden with palm oil nuts) pulled into the side of the road as the call to prayer sounded out from the tinny speakers of the aluminium-roofed mosque.

Already soaked in sweat and bothered by a legion of mosquitoes and sweat bees, our desire to leave this place and be swallowed by the forest was overwhelming. We carried our kit bags on alarmingly bendy planks over the fly-blown mud and stacked them inside a train of long-tail wooden boats. Their engines sounded like an over-revved moped with a hole in the exhaust. We chuntered through the grimy waters hunkered under ponchos against the constant rain, fingers in our ears for the engine roar. Knowing we had two weeks more of this ahead, Aldo and I barely spoke for the entire journey.

Just a few hours upriver, though, and the thrashed-up mess of secondary forest started to evolve into something prettier. No big mature trees – those had been taken long ago – but plenty of dangling lianas and drooping palms over the brown waters. There were also an unusual number of monkeys. To begin with, my new friend Pak Udau, who was doing the steering in the front of our boat, would point to the monkeys from a distance, and I'd squint into the trees before finally seeing them. He seemed to be some kind of monkey-spotting shaman! After a while, however, I realized that in one particular flowering tree there were *always* monkeys – long-tailed macaques, sitting together in extended family groups, grooming each other and watching us intently. If we slowed, they'd leap in panic from their boughs, stampeding off through the canopy into the distance.

As the dusk drew in, we saw high above us some more orange-looking primates. Looking through my binoculars I could see the retroussé nose and orange flat-top haircut of proboscis monkeys. I didn't see any adult males. You can't miss them; they have a nose like a pendulous breast.

Our roaring long-tailed motorboats were too loud to have a conversation over, but I still heard Aldo gasp as we rounded a corner.

'That is seriously impressive!'

Beyond the bend, the green ripped skywards in vertical rock-walls hundreds of metres high. The limestone cliff was in white and orange, but with stains of black running through it, as if someone had spilled a pot of ink somewhere up on top. The rock faces were highly detailed, carved and sculpted by rainfall. And already we could see plentiful evidence of the caves we'd come here to find. Finally, I allowed myself a spark of excitement.

As we would be travelling for at least two and possibly three days by boat, and would then have access to porters, I had made a very clever decision. Instead of following my own rules and packing light, I had brought a few luxuries. Three changes of clothing. Underwear. A towel. Before the trip I'd gone to the poshest supermarket I could find and bought gourmet soups, hard cheeses and quality chocolate. I'd brought both a hammock *and* a small tent, a sheet *and* a sleeping bag, even a portable speaker for music. I was feeling so, so smug . . . right up until that first night of jungle camp. Then, as I scrabbled around in a thundering rain storm, trying to find the essentials I needed to put up my hammock, I remembered that going light is not just about the weight you have to carry. It's about streamlining. Perfecting your system so that there is nothing extraneous. By bringing the equivalent of the kitchen sink, I had completely compromised my own system. It was a disaster. There were too many little bags inside too many big bags. In the dark, I couldn't find my spoon, my earplugs, my bug spray or anything, in the hammering rain under a tiny tarpaulin.

Next morning, I packed all my luxuries into a bag, and left it behind (after first shovelling all the chocolate into my face).

From that first camp onwards, it felt like the Borneo I dreamt of when I was a kid. Giant trees hung with vines twisted around each other like plaits. Holes, hollows and the crooks of branches were bursting with orchids and ferns. Squadrons of hornbills flew over our boats in single file, their wings sounding like the whoop-whoop of a helicopter firing up its props. Occasionally, stork-billed kingfisher would flash in iridescent colours up the streamway ahead of us. They were the size of a pigeon and bore a huge comedy orange beak that seemed too big for their body! And then there'd be a gap in the green ceiling, and we'd see blue sky and, looming over us, the massifs of the karst.

Local people call these mountains *Batu Gergaji* – the sawtooth mountains – for good reason. The peaks run in lines, with each summit jutting upwards like a worn canine tooth. The karst dominates down at river level too, with overhanging limestone looming over us as we pass. In some places the effect was so dramatic that, as our boats were gunned through a cave, we had to lean to one side to avoid the huge stalagmites dripping water from their ends down into the river.

At one point we rounded a corner to see a lashing body and a thrashing tail; a saltwater crocodile rampaged off the bank and into the drink. It was at least three metres long and plenty big enough to take a person. We'd have to think carefully about bathing in these murky waters.

That evening we made camp in a bird nest collectors' encampment. It was nothing more than a cleared piece of forest the size of a tennis court with a few chunks of blue tarpaulin tied up with thin vines. Our Indonesian contingent set about making it home, everyone sporting spindly moustaches and dangling scented clove cigarettes called *kretek* from their lower lips.

Indonesian is the language I speak best other than my own, which is the greatest gift for making fast friends. Sadly, it had been a couple of years since I'd spoken it, so I inadvertently kept peppering my Indonesian with Spanish, French and Japanese words. Our new friends were too polite to comment, and never pointed out that I was talking complete gibberish. I spoke to one of the younger men, Pak Setiadi, who stood to my shoulder height, but could have danced through the steepest jungle terrain carrying me on his back. He had smiley eyes, and gently coaxed the others to speak slowly so I'd understand. I liked him instantly.

'What do you usually do for a living, Pak?' I asked.

'I collect the birds' nest,' he said, 'and I have them in my house.' He meant the artificial swift nesting towers you see everywhere in this part of Kalimantan. They have speakers inside playing ear-splitting swift calls to attract potential swiftlets to nest.

'Is it a good living?'

He pursed and lips and shook his head. '*N'dak* – not so much. It used to be that you could earn 50 million rupiah from one cave with the nests,' he said, 'but now the price has gone right down.'

This was the first I'd heard of this. Swiftlet nests are the ingredient in the Chinese delicacy bird's nest soup. The nests, which are made by the birds out of spongy dried saliva, are collected, stewed and sold for preposterous prices. The search for nests has driven the exploration and exploitation of caves throughout Southeast Asia. At one point, nests were considered pound for pound the most valuable natural commodity on earth.

'Is that difficult?'

'Yes, of course,' he replied. 'But at least now it's more safe. Before it was the Wild West up here, people have guns. If you come to the cave they will think you are a robber and shoot you. *Berbahaya sekali* – very dangerous.'

I've seen collectors scaling spindly bamboo ladders hundreds of metres up into the roofs of caves to collect the nests. No harnesses, helmets or safety ropes, just barefoot climbing to the height of Big Ben. In Java they dangle on ropes over crashing seas and will be deliberately swept into tidal alcoves to find them. The men willing to risk the most are the ones who get the reward. And here, they risk all-out jungle warfare, purely to collect a sackful of bird spit.

I smiled to show I understood, and patted him on the shoulder. '*Gracias, señor.*'

From our next camp the river became a stream and then a trickle. We battled our way up, sometimes using the engines, more often wading waist-deep through the water, hacking away with machetes at the branches that covered the paths. The long-tailed motors were little more than a three-metre-long metal pipe attached to the engine, with a propeller at the other end. At one point, our driver Pak Stefanus hit a rock with his prop and it span off, flying into the stream like a lethal frisbee.

'*Sudah rusak,*' I said to Pak Udau. 'It's broken!'

'*Tak perhah rusak,*' he replied. 'It's never broken.'

His older brother Pak Stefanus fixed it back in place with a bent nail, a few whacks with a rock, and some winds of green reed he'd plucked from the riverside.

'It's an unwieldy appendage, isn't it?' Aldo commented of the long tail.

'You're an unwieldy appendage,' I replied.

It was a long day, with over nine hours of dragging, hacking, puffing and cursing. At one point we came to a tree trunk with the diameter of a dustbin lid that was downed across the river. Just as we were all sighing with dismay, Pak Stefanus reached under the blue tarp at the back of the boat and pulled out a giant chainsaw. We all had to turn away while he set to with the tree, as he was stood barefoot on top of the dead trunk with the chainsaw churning away in the water, sending spray everywhere. You had to admire his skill with the chainsaw, though. I would have taken off both my feet and probably electrocuted myself.

The sun was long gone and the rain had started to hammer down on us by the time we came to the end of the river. Brown water emerged from a cave in a clammy green rock face, a textbook 'resurgence' where water bursts out from the rock, and a spectacular source for our river. Pak Pindi had been into the cave beyond before, but said it remained unmapped, unsurveyed. That seemed like an obvious first objective for the morning. For now though, we needed to get our hammocks up and our heads down.

The camp was clearly a great spot for wildlife, mostly obvious from the fact that within an hour of us being there the locals had three mouse deer barbequing on the fire. (Imagine a muntjac deer the size of a corgi and you've pretty much got it.) Once it was dark, I went down to the river and got in for a bath. By now the water had cleared right up, and you could see the bottom. Almost immediately, I leapt what felt like six feet in the air and shouted something unprintable. A giant snake the size of my arm had swum right over the top of me. But when I focused my light on it I realized it wasn't a snake but an eel, olive green and covered in white spots. And in its mouth was a fish it had just caught, which was the size of a dinner plate. I shouted up at Aldo to come down and look, but before he could get over and see it, one of our local friends had slid down the slope into the water, machete in hand, and taken the eel's head off with one almighty chop.

I stood there watching the wildlife we'd come to see being dropped into a pot of boiling water for our porter's dinner and felt sad, cross and conflicted all at once. We'd made it very clear from the start that there was to be no hunting on this trip. There were no guns or fishing rods allowed, but everyone needed to have a machete, and that was enough. Hunting is a way of life for our Bornean friends.

Luckily, they weren't interested in eating the bugs, and in Borneo you get the most dramatic on earth. Many of the trees are covered in fulgorid or lantern bugs. These weird critters have giant wings covering an orange and green speckled body, and then a preposterous rostrum or snout above their eyes, curved up into the shape of a Belisha beacon. Early naturalists believed they had a light at the end of the snout, hence the name. We also came across some mighty stick insects, the largest of which was a Bornean giant stick, huge, black and lobster-like, menacingly wandering over my hands with its conspicuous spiky armour, looking for all the world like a devourer of babies rather than a harmless leaf eater.

Next morning, we waded into the darkness of the resurgence cave, following the course of the water. Once inside, there were a few possible choices of direction, but one tunnel had an almighty wind blowing out of it, a sure sign of bigger passages beyond. Here in the first few chambers and tunnels, we found cave-adapted wildlife, such as huge scutigera centipedes with straggly legs they use to sense their prey, having evolved to function much like tactile antenna. These are some of the creepiest creatures on earth, moving like a scuttling nightmare across the rock. They are also venomous.

'People say of this animal, it is the message home,' Pak Pindi told me.

'What do you mean?' I asked. 'It's like a spirit animal?'

'No.' He shook his head. 'If you get bitten by it, you should then send a last message to your family.'

'Oh,' I said. Not much I could add to that.

'One of the caves here has an image of this scutigera on the wall,' he said.

'Really?' I was astonished. 'I've never heard of a representation of an invertebrate in cave art before.'

'I think this is the only one.' Pindi said it as if he were telling me what he had eaten for breakfast. I guess it must be difficult to impress someone who knows so much and has discovered things beyond the imagination of most scientists.

The cave was full of circling bats of several different species, with the swiftlets inevitably out foraging during the day. Cave racer snakes with striped tails and bodies glinted gunmetal grey in our torchlight. Come dusk, they'd be dangling down into the air, snatching bats as they came past on the wing. I've spent many miserable days trying to film this spectacle, sitting on my own in the pitch-dark (they're put off by any hint of light) with an infra-red camera. I only managed to see it happen a single time, clearly triggered by the vibrations of the bat's wings. Once caught, the prey was briefly constricted, before the snake steered the bat around so its head went down its gullet first. The snake then chugged it whole, which looked rather like a toddler trying to swallow a beach ball.

The cave opened out into a dramatic chamber, with flowstone chandeliers pouring down one wall. On the other wall a stalactite hung down, missing its end. Water was pouring out of it like a power shower, dropping down four or five metres into the water.

'You should do us all a favour and have a proper shower, Aldo,' I quipped.

'It's you needs a shower,' he retorted.

'Snappy comeback.' I put on the voice of a mocking teenager: 'I don't smell, you smell.'

Aldo backlit the shower with his head torch. As the water flooded a basin below, droplets bounced up in the light like fireflies.

'I reckon I should get one of these put into my new house,' I mused.

'What, a shower?' Aldo said, 'Most people have those nowadays, dude.' Dammit. Backshall one, Kane two.

The real spectacle, though, was yet to come. As we rounded a corner, we could see natural light spilling down. It was a doline, a sinkhole, where the roof had collapsed, allowing sunlight to pour down into the darkness. As with all things, contrast increases the drama. After pitch darkness, the shafts of sunlight breaking into the cavern were celestial.

The next day was set to be our first big trek, heading up and over a high pass on the ridgeline that runs north to south and forms the sawtooth mountains that give the area its name. Pak Ham, one of the main bird-nest collectors, led the way. Ham had been described to me as a forest hermit, a man who had been wandering the forests for decades, finding obscure caves and collecting birds' nests. Pindi had been relying on him as a source of information ever since the 1990s. Ham looked ten years older than me, was wearing a pair of two-dollar plastic-studded football boots without laces, and looked as if he were strolling. Within an hour, I was gushing sweat and the team looked as if they had been hosed down. Ham's pace remained relentless, never resting, never slowing.

Then the way pitched upwards, and we picked our way over steep spiky karst, with terrestrial vines that seemed designed to hook your feet. As we climbed, we saw orangutan nests – painstakingly woven clumps of leaves in the crook of a tree. Orangs make their nests daily and move on. One we found had super fresh droppings nearby, suggesting we had missed our man of the forest by a matter of days.

I nearly stepped on a snake – a striped keelback – which slipped across my path. As I picked it up to show it to the crew and talk about how harmless it was, it sunk its teeth repeatedly into my hands. Not wanting to highlight my mistake, I walked on fast, giving the copious leaking blood time to congeal. The floods of blood from tiny wounds suggested an anti-clogging factor in the snake's mild venom.

Eventually, after five hours of hot slog, we came to a teeny pass between two bobbly peaks. The gap in the limestone was so narrow I had to take my pack off and turn sideways to ease through. The jungle is by its very nature a claustrophobic place, a green mausoleum where sounds are deadened, where sunlight and wind rarely penetrate, and where views are few. Here, from this high pass 400 metres up, we got our first proper view of Sangkulirang, after five days swallowed up by the forest.

I had been mentally preparing myself for this moment since before I left home – the moment when we would see out to the horizon for the first time, would see the limits of the karst and the encroaching plantations beyond. I knew it was going to be a time for solemn reflection, for realizing that this perfect forest is under siege, its days numbered unless we spring into action right now.

But that was not what we saw. Instead, we looked out to the horizon and saw nothing but mountains and jungle. To the south, the green dropped down to lowland forest. To the north was just the mountains. Eastwards, a high ridgeline obscured our view, but to the west it was uninterrupted, perfect, no sign whatsoever of humans. When we put the drone up into the air to get a broader sense of place, we saw our own ridgeline was a stegosaurus's back with jutting olive plates. Below them were bizarre half-oval outcrops, looking like the alien's eggs in her subterranean birthing chamber. It was the first time in 20 years that I had looked down from on high anywhere in Borneo without utter despair. This was the hope I had lost, proof that there is still so much here worth saving.

And from there the good got sublime. We descended along a natural balcony until we came to a small innocuous cave. Pak Ham darted inside (to have a sneaky peak for a bird's nest and a quick buck, I thought) but then returned just minutes later and beckoned me in. We followed him through a dark passage that curled between cave features before opening out into the mouth of another opening. Again, the views out were spectacular, but Ham beckoned us on again. Barely stopping to soak in the last of the sunlight, Pak Pindi and I ducked under a small overhang and into a giant cathedral room, lit by sunlight from two gaping exits. Instantly, I saw pigment on the walls that was clearly primitive cave art, but Pindi wasn't stopping – he had something special to show me. We wound down past the cavern's main feature – a dark green stalactite like a column from the Parthenon – up past a flat sandy floor where we would eventually make camp, and into a smaller gallery. By now my eyes were catching stencils, etchings and daubs of pigment all over the place; I had to force myself not to stop.

'There!' Pak Pindi said with a flourish, 'this is my favourite.'

He was pointing to two sets of hand stencils maybe ten metres above my head. There were six hands that had been blown around with a dark purple pigment and were as clear as if they had been done yesterday. Around those were many more, fainter prints in orange.

'The purple one is 20,000 years old,' he said. 'The orange is much older, at least 40,000.'

My breath caught in the back of my throat. A wave of something tangible swept over me. Forty thousand years. The oldest cave art I'd seen before was the Lascaux caves in the Dordogne in France. They might be 20,000 years old, daubed back when sabre-toothed cats and woolly mammoths roamed the north. Yet here were crystal-clear prints from way before then. As we continued around the cave, Pindi showed me countless images, any one of which an archaeologist in Europe would have made their life's work. Yet here, shrouded by the forest, this whole cave had only been seen by a handful of outsiders. We were the first to commit the images to film.

There was an extravagant shield-like lizard, a turtle, a map showing a river and an estuary, human figures and a banteng (a kind of wild bovid/cow) that is believed to be the oldest figurative art ever found. But it was the hands that grabbed your attention. Hundreds of hand stencils, everywhere you looked. If you followed the rock walls up the most tenuous of rock climbs, at the top would be the hands. In every spot where moonlight or dawn sun would fall, hands. The representations were clustered together but, even if separated by 20,000 years, they never overlapped. The people coming through here late in the ice age respected the markings of their ancestors just as we would.

The whole team – both Bornean and British – wandered around speaking in hushed tones. There was a respect, a reverie, to our every action here. It felt somehow both a heresy and an honour to lay out our beds under the timeless etchings. Every move we made was done in silent contemplation of ancient ancestors who had done the same things. It is the closest I've ever had to a religious experience.

As we unrolled our sleeping mats in the flat dusty centre of the cave, we realized another reason why these caves must have been so important. Four hundred metres of elevation on the ridgeline meant there was a constant cooling breeze which kept away the mosquitoes. After the unremitting sweat of the jungle, it was the absolute perfect temperature. Without even a sheet, it was the most comfortable and soundest night's sleep I'd enjoyed through this entire year – including all the fanciest hotels, even better than my own bed, back home in England. We had protection from the elements, a soft bed, a view; timeless attractions that must have been as valuable for our ancestors as they were for us.

When we woke at dawn, we all wandered up to the viewpoint to watch the sun come up. As it crested the ridgeline in front of me, gibbons started singing to each other, their lunatic whooping calls building to a haunting climax. Tendrils of mist billowed in the valley below me, as the forest sighed. It was impossible not to imagine our forebears clustered here in this very spot, looking down into the valleys for water, game and the route ahead. They must have felt blessed by the gods here. As did I.

Expectation and context are everything. I had been determined that this trip would be just a slog, a headache and a misery. That I would be beset by missing a month of my little boy's life, that I would hate every second. I had already written the Sangkulirang story, and it was a cautionary tale of a once-perfect place under siege. The trip would be important, but ultimately unfulfilling. So completely blindsided was I by the reality that my reaction to it was more emotional than I ever thought possible. I wandered round on the brink of tears for the entire two days we were in the secret cave. And it wasn't just me. I saw other members of the crew on their own, placing their own hands in the prints of prehistory, silent, teary, even weeping.

Sitting out at the mouth of the cave as the sun rose and set was the first time in the entire epic year that I was truly content. I allowed myself to realize we had done it. We had achieved everything we wanted, and more. At night, we sat around a small fire and recounted our favourite campsites of the year. For Aldo, it was the high glacier camp in Greenland. For Parker, our huddle at the top of the undropped falls in Oman, knowing what the next day had in store. Cameraman Mark bizarrely chose the sweat-fest hovel at the top of Backshall's Backdoor in Mexico, proving there is no accounting for taste. But for me, there was no contest. This was not just my best camp of the year, but of my life. A time to put everything in its right place. For the first time, I considered the real possibility that this would not be my last expedition, that maybe there were a few years left in me yet.

That night, the nearly full moon cast its peeking eyes into our cave, washing the walls with an ethereal light. The main walls where the art was found were most brightly lit. Pindi explained that all the caves he had discovered here were aligned east to west, so they caught the best glow of the sunrise and sunset, and were additionally illuminated by the full moon.

The following morning we were again up before the dawn, summoned by the whoops of the gibbons. At the cave mouth we marvelled at a growing inversion below us. The billowing clouds down in the valley coalesced into a blanket of white.

As I left the cave, Pak Ham showed me where he had inscribed his name near the exit the first time he had come here in 2000. I was surprised that anyone would gouge their graffiti into the rock. But Pindi pointed out that the art of Sangkulirang was essentially the same thing. It is leaving your mark. Your own tag. It is seeking immortality, saying 'I was here.' And what could be more universally human than that?

With our expectations thoroughly exceeded, it was with renewed excitement that we set off on the exploratory stage of the expedition. From here, we were set to head north, past the ridgeline we were now on, and to an entirely separate limestone massif. Pak Ham said he knew of some caves high up on the rock face that had not yet been explored for their cave art, and were little known even by the bird's nest collectors. He was super cagey talking about them while certain other members of the team were around, clearly fearing his precious territory could be usurped. Although we were desperate to talk to him about the leads, we had to wait until we had him alone before we could broach the topic.

Sitting around a crackling fire, with the blackened kettle whistling and shadows creating an Indonesian shadow puppet display on the cave walls, Pindi had the opportunity to quiz Ham about what he had seen. Even he was only hearing of them for the first time, and he was brimming with excitement.

'He says the caves are high up,' Pindi said. 'These are always the most sacred places; most of the best caves are the highest.'

We nodded, encouraging him to ask more. Ham took a long draw of his cigarette and cast his eyes around as if some spy might be waiting in the shadows, then spoke again, cautious, conspiratorial.

'And he says he saw some marks. He is not sure if they are paintings.' At this Pindi pursed his lips. 'Hmmm . . . this is not so good. Usually if Ham says he is sure, then there is art. If he is so-so, then sometimes it is just natural marks.'

At this we all exchanged worried looks. Pindi had already told us that only one in ten caves here had given him results. That was a lot of slog to find nothing at all. Aldo summed it up best: 'There is no helicopter rescue from here. No way out other than from where we've come. Someone breaks an ankle up there to the north, it'll take a week to carry them out on a stretcher.' And then, with classic Aldo bluntness, 'So my only safety advice is: don't get injured.'

The next six days were pure jungle hell. We hacked a trail through virgin forest, with colossal rainforest trees towering above us, the hoots of gibbons resounding round the steaming dark forest floor. The leeches were relentless. At every rest I would reach down and pluck up to 20 of them off my boots and trousers. Inevitably, there would be four or five attached to a black blood seeping wound, fat and full, gorged on my blood. At one sitting, I flicked off my unwelcome guests then got lost in conversation. A few minutes later, I looked down to see an army of leeches inching towards me in a line, like something out of a horror movie.

The bees and mosquitos sought out any spare inch of flesh; stinging trees and hairy caterpillars left welts that would itch for weeks afterwards. I also got my worst ever case of chiggers. These delightful burrowing mites attached themselves to my backside – probably when I sat on a rotten log – and laid their eggs into the flesh of my buttocks and belt line. The itching was beyond furious.

All of this, though, was nothing to the rattan. Rattan is a climbing liana, which is the plant behind the phrase: 'don't fight the jungle'. A thin, tough and usually springy vine, it is used by locals as a natural string. It makes its way through the forest canopy using a series of tiny hooks along its length. While hiking through the forest it's relatively easy to stop and untangle yourself, but when moving up or down steep slopes, it takes a sinister turn. The rattan attaches itself to you, and then, as you step up or down, rips through your skin. It's totally unbreakable. After one particular descent I looked as if I'd been for a short back and sides at Sweeney Todd's.

We slithered up and down muddy, slippery slopes, only managing to keep our feet by grabbing hold of roots and spindly trunks. Our route was taking us up the valley, heading ever north. From up on our high gallery, the trees below us had seemed like matchsticks. However, this was actually the best primary forest I've seen in Borneo (outside of protected national parks). The locals call primary forest (one that's never been logged in any way) like this *hutan raja* – royal forest. Borneo has the tallest tropical trees on earth, with the record being a tree over 100 metres tall that weighs more than a Boeing 737. Minus roots. That's double the height of Nelson's Column.

Down here in this sacred valley, the trees had to be seen to be believed. They were giant buttress-rooted monsters. If our entire team tried to stand around them hand-to-hand, we'd all have our faces pressed to the bark. It was a Tolkienesque jungle of absurd profundity, though with the most open and appealing forest floor. This architecture – giant trees with big gaps between them all, and little down on the ground – is believed to be responsible for the fact that so many odd animals have evolved to 'fly' here. There are flying lemurs,* squirrels, geckos, frogs, even a snake which flattens out its ribcage to form an aerofoil and 'swims' through the air.

At night, I sat in the locals' blue tarp camp, drinking coffee so sweet it would strip the enamel off your teeth, and chatted to them about their forest and their place in it. It was humbling to see how they managed to create a home in a matter of hours out of vines and branches. Pak Satriadi carved a machete handle out of wood

---

* Not actually related even closely to the lemurs of Madagascar, but unique gliding mammals called colugos.

almost absent-mindedly in about an hour while we talked about forest wildlife. He told me about the biggest snake he'd ever seen here (a 6.7-metre, 50-kilogram reticulated python, which is specific and probable enough to be accurate), and the king cobras he'd seen nesting nearby. It seemed the bird's nest collectors had been oblivious to the meaning of the hands in the caves until they had met Pindi, and certainly had no idea of the age of the art. Now though, they treated them with the same reverie as we all did. When I told them all in my clumsy Indonesian that I thought our last camp the most beautiful cave in the world, they erupted into spontaneous applause, clearly proud of their own little gem.

Our biggest and most challenging day was up to the cave that sounded most exciting to Pak Pindi. The last section was a completely vertical climb, up jagged limestone in a downpour. We were already shattered, having slogged uphill for five hours just to get close. Add the remoteness, the total impossibility of any help or rescue if anything had gone wrong . . . it was terrifying. And worse than that, when we finally reached the caves high above, they yielded nothing more than a few smears of possible paint. We retreated to a camp in a miserable damp valley cave, beaten and humbled.

It was late on one searing afternoon that we broke out to a view of quite staggering majesty. Off to the north of us were two more completely new mountain ranges, all smothered in forest. One featured a peak that looked like the Matterhorn, just bursting up out of a curly kale salad. A giant hornbill cronking like a lost goose landed in an emergent tree right in front of us, eyeing our party with an alien glassy eye. And before us was another cave to explore.

The beginning was inauspicious: a duck into a dark chamber, and then a fumble around through several hundred metres of total black, up to our ankles in swift and bat guano, as bats thundered blindly into our faces (I know they are not actually blind, but that didn't seem to stop them!) and furious bees stung every open inch of skin.

'We don't find art in dark caves here in Sangkulirang,' Pindi said. 'There is usually natural light.'

Eventually, though, we did break through into a chamber that was dimly lit. At the far wall, two window-sized skylights provided a vague snooze light to the room.

'There!' shouted Pindi with glee. It was a hand, emblazoned with orange, strings of deliberate dots across the palm.

'It is the old kind,' he said. 'More than 40,000! And there also!'

Above him was a single huge handprint, the thumb alone like a healthy banana.

'Look at the size of it!' I replied. 'He'd have made the world's greatest-ever hitchhiker!'

Seconds later, and the whole team were shouting and pointing, as new discoveries appeared all over the walls and ceiling. There was an etching, inscribed deep into the limestone in the shape of a deer.

'It is the first we have ever seen in Sangkulirang!' Pindi said. 'And these, these handprints, these are also the oldest ones!'

Pindi then studied the structure of the room. 'These windows in the rock,' he said. 'First thing in the morning they would have light shining in through them, like a spotlight.'

We turned and looked at the walls that would have been illuminated. All over them were more perfect handprints. After dawn they would have been lit up in the dark room, like paintings in an art gallery, highlighted to focus on their beauty.

Lying on my back on the rock, I found two images on a part of the ceiling that made me choke up. They were tiny hands, emblazoned in black – not the standard colours, and not stencils, but handprints. A baby's hands, lathered in paint, then pressed up here onto the ceiling.

'This is the first positive handprint we ever saw here,' Pindi marvelled again, trying to make sense of this unique symbol. To him, every print was more than just a tag, it was a story to be unravelled. To me, it was obvious. It looked just like the plates and cups Helen and I had made that Christmas for our families, with Logan's tiny hands printed on them. We wanted to give everyone a memory of our boy, to marvel at his infant doll-like paws, knowing that all too soon they would no longer be such miracles in miniature.

The art here connects you to our forefathers in ways impossible to describe. You can feel the heartbeats of the past here. Those people – possibly clad in the furs of long since extinct animals – they appreciated a good view just like we do. They needed shelter and food, and when they had them, they wanted to express themselves, to leave their mark on the world, just as we do. And they valued their families, they loved their babies just as I love mine. A family had lain here in the same dirt as me, when much of my homeland was still in the midst of an ice age. They had taken their baby and placed their hands up here. Maybe he had struggled just like Logan did, but they persisted. They wanted their baby's mark to be left here for posterity.

Or maybe they didn't struggle. Infant mortality was rife; maybe it was a cold hand placed here, a simple epitaph. Perhaps black had the same meaning to them. Were they mourning? Did their tears roll down the limestone as mine would? At this, the dust caught in the back of my throat and I was too overcome to speak anymore.

EPILOGUE

Dear Logan,

Your first year was memorable for so many reasons. In that one remarkable year I had the honour to walk in places where no human had ever stood, and in places where they have walked since the dawn of time. I faced my demons, and was party to discoveries I never dreamt possible. Yet my biggest learning experience was simply becoming a dad. The thing that changed my perspective more than any archaeological discovery, near-death experience or victory was holding you, my bloodied baby, in my hands for the first time.

I have always been searching for something. A selfish swine who spanned the globe looking for the meaning of life, only to find it in the most obvious of places, in the same way people have been doing for tens of thousands of years. You are my biggest adventure, the thing that makes me more proud, more excited, more protective than I could have even imagined just a year ago. But you also make me anxious. I want to bundle you up in bubblewrap, to cushion you from the very bruises you need to grow and flourish. I fear for you constantly, for your safety, for your future. And that's where the biggest change has come. I always patted myself on the back for being a decent sort of chap, for wanting to make the world a better place and all that. But now it's clear I have been found wanting. Me and my generation have been asleep at the wheel.

For tens of thousands of years, we humans were just another animal, struggling in our world alongside every other species. And now, within a few decades, my kind have turned into an all-consuming ravenous monster. The more we are the master of our world, the less we are connected to it. The wealthier and more comfortable we become, the less we are prepared to sacrifice that comfort for a better future. We are passing on a poorly planet to your generation, in a way that makes me deeply ashamed. I have nightmares of a First World War propaganda poster, with you sitting on my knee, asking, 'What did you do in the Great Extinction, Daddy?'

In this one year, I've seen the world through the lens of a father for the first time. The forests I want to wander with you are being strip-mined for their resources, the oceans where I want to teach you to swim and dive pillaged and polluted. Arctic ice may soon be no more. In my lifetime the world's population has doubled, our forests reduced by half. Great gyres of plastic choke our oceans, 50 per cent of the animals on earth have gone. I fear what will be left for your child or grandchild. The human race is at the helm of a sixth extinction event, one greater than any our planet has seen before. Greater than asteroids, more powerful than tsunami, more destructive than any volcano.

But I've also seen so much that's worth fighting for. Wild places, a paradise for people and nature alike. Sanctuaries where you can be alone with friends, where you can stare up at a smog-free night sky as our ancestors did, watching shooting stars streak across the firmament, wondering about our place in it all. There are still millions of species left for you to discover, thousands of mountains left to climb, rivers to run and dark places to illuminate . . . so much still to know and understand. I want you to see all of it, to have expeditions of your own, big and small. There is plenty of wonder left in this world, and the time for protecting it is long, long overdue. Now we just have to figure out what to do. How to stand up for this one, magnificent blue planet so we don't destroy it before we have even begun to understand it.

Love,
Dad

ACKNOWLEDGEMENTS

One wild year. Seven nations, around 120 nights under canvas (well, actually under dripping tarpaulins, cave ceilings and the night sky – never actual canvas!). Countless miles trekked, paddled, skied, climbed and even crawled! We covered a lot of ground, and did our best to behave in ways we could all be proud of. UKTV, True To Nature and I knew we couldn't set out to make *Expedition* in good conscience unless we took steps to ensure the entire production would be carbon neutral. We are working with Albert and Natural Capital Partners, and will announce more details as soon as we can. Additionally, I offset my own flights through forest purchase systems. We endeavoured to take all our rubbish out of the field with us and recycle it. Efforts were made to involve local people in all of our projects, as key members of our core team. Though we asked permission to name our Arctic peak and desert drop after Logan, the two new caves in Sangkulirang were (after discussion with our Indonesian brethren) named Gua Satriadi and Gua Rusti, after our friends Pak Satriadi and Pak Rusti who led the way to them.

We owe a debt of gratitude to Pindi Setiawan, Robbie Schmittner, Khaled Abdul Malak, Michel Boeijen, the elders of Ittoqqortoormiit,

Darren Clarkson-King, Justin Halls, Khalid al Hikmani and Guillermo de Anda for their knowledge and experience. This would have
been impossible without them.

Nothing was 'staged' for the cameras. The crews did their very
best to record the expeditions happening for real in real time. John
Livesey deserves credit for his revolutionary camera systems that
could film constantly, capturing every eventuality. Parker Brown
likewise, for flying drones while we got on with expedition business!
The nature of television means you cannot show the footage back in
its entirety, and editing is inevitably somewhat subjective and select-
ive. However, the whole team is proud of the results on screen, and
happy that they represent the spirit of all our adventures.

Huge credit goes to my producers James, Rosie, Rowan, Ali
and Sanna, who, along with Anna, Briony, Annie and Joff, worked
tirelessly both in the field and in the darkened room of the edits to
make the programmes work. The True To Nature Production team
worked tirelessly back in the office, making the impossible happen;
a huge thanks to all of you. My camera operators who managed to
get through the expeditions while operating a hefty 4K camera! To
Katy, Mark, Graham, James, Keith and Stu, I salute you! Likewise
to Parker and Nick, as we were never 'waiting for sound'.

Every effort was undertaken to ensure the veracity of our 'firsts',
and to make sure they were registered afterwards with the correct
authorities. That is not to say that none of these achievements will
ever be challenged; it is the nature of the adventure game that this
may be the case. We have already made contact with a Dutch

mining company that ran transects on the Sand Creek in Suriname in the 1970s, who say these probably ran through part of the creek we named YiYi. They do not lay any claim to have seen Ile Falls, to have descended the river, or to have created any record or map of the river. The scientific method of peer review applies equally to any expedition, and should anyone in future claim to have preceded our efforts, I welcome their comment.

Due to the nature of publishing and television schedules, this book had to be delivered before I got on the flight to head home from Borneo. It is more raw than I would have liked, taken directly from the diaries I wrote daily whilst in the field. I hope that the lack of polish in the text will be countered by a real sense of how it feels day-by-day to be on voyages like these. Thank you to my editor Albert for being honest with me, and helping me make this a book I can be proud of. Likewise to my literary agent Julian, who has been there with wise words ever since my very first book.

Sanna and Jonny deserve my gratitude for managing to put together films that are destined for screens in well over a hundred countries worldwide. And to my tireless boss/partner in crime, the indomitable Wendy Darke, who was the master behind making this long-held dream of mine a reality. Wendy, you are a force of nature, and I would vote for you as prime minister any day!

And lastly, to any budding explorer who would like to follow in our footsteps, I cannot urge you enough to chase those dreams! The complete first descent of the Chamkhar Chhu is still out there, and must still rate as one of the world's greatest remaining unrun

rivers. Yucatan will be providing new cave diving finds for generations to come. The tropical forests are alive with rivers awaiting
exploration, and the Arctic still has more peaks unclimbed than
climbed. Satellite imagery puts discovery at the reach of anyone
with an internet connection. Grants, new media, social media and
obscure funding are making ambitious dreams a reality for budding explorers every single day.

Don't be afraid to start close to home – my first expedition was
probably trekking up some Welsh mountain with Mum and Dad;
my first kayak expeditions were on the humble Wye with the Scouts
and the Duke of Edinburgh award scheme. It's essential to learn the
tools of the trade, to be capable of making your own decisions, being
self-reliant in environments that can turn uncomfortable quickly.

There are many good reasons to get involved in adventures,
however big or small. In a world where loneliness and alienation are
rife, the friendships you forge in the field are beyond value. The
benefits of fresh air, sunlight, exercise, a connection to nature and
healthy 'clean' food are well documented. I have several people
very close to me who suffer from dementia and/or depression;
these lifestyle elements are all part of the mainstream prescribed
treatment for both. More than anything, we need to rediscover and
reconnect with nature on a grand scale. Jacques Cousteau said:
'People protect what they love.' Helping to show the world its own
beauty, sharing your passion, giving others a reason to care – all of
these things can be a part of the conservation solution.

INDEX

SB indicates Steve Backshall.

INDEX

INDEX

INDEX

· · · · · · · · ·

INDEX

· · · · · · · · ·

INDEX

INDEX

• • • • • • • • •